PARASITIC DISEASES OF BUFFALO IN INDIA

PARASITIC DISEASES OF BUFFALO IN INDIA

B.B. BHATIA, M.V. Sc., Ph. D.,
Professor & Head, Department of Parasitology,
College of Veterinary Sciences,
G.B. Pant University of Agriculture &Technology,
Pantnagar (Nainital) UP.

AND

P.P.S. CHAUHAN, B. Sc., M.V. Sc., Ph. D.,
Research Officer (Parasitology),
College of Veterinary Sciences &
Animal Husbandry, C.S.A. University of
Agriculture & Technology,
Mathura, U.P.

A FALCON BOOK From COSMO PUBLICATIONS

NEW DELHI INDIA

1984

A FALCON BOOK
From COSMO
PUBLICATIONS

PARASITIC DISEASES OF BUFFALO IN INDIA

First published 1984

Published by
Rani Kapoor (Mrs)
For FALCON BOOKS
Parent Publishers
COSMO PUBLICATIONS
24-B, Ansari Road,
Daryaganj, New Delhi

Printed at :
Swatantra Bharat Press, Esplanade Road,
Delhi-110006

FOREWORD

The authors have brought out a concise but vivid description of important parasitic diseases of buffaloes. Buffaloes are important dairy animals in our country contributing more than half the total milk production. We have 67 million of the world's 126 million population of buffaloes, which produce about 13 thousand tonnes of milk annually in India. The best breeds of buffaloes are found in the Asian sub-continent. In order that their optimum genetic potential of milk production can be utilized, their management, nutrition and health have to be improved.

Although it is generally realised that parasitic load of our animals reduces their productivity to a large extent, the specific information about the type of parasites, their effects and mode of infection is not available. Parasitic diseases of buffaloes are many and of varied types. These cause substantial economic loss to the dairy industry. Though buffalo and cattle share mostly identical infections, there are some specific problems, which need to be understood.

I am confident that this book, a first attempt, would prove useful to veterinarians. It is, of course, expected that the veterinarians engaged in the management of buffalo herd would benefit the authors by giving information and suggestions for bringing out a more informative and illustrative next addition.

Dr. I.P. Singh, M.S., Ph. D.,
Dean, College of Vety. Sciences,
Pro-Vice-Chancellor,
G.B. Pant University of Agriculture
& Technology, Pantnagar Nainital.

PREFACE

The scope of Veterinary Parasitology is increasing day by day with the continuous additions of new parasites and modern knowledge of their morphology, physiology, biology and of the host response to them. With the coverage on newer hosts in the purview of Veterinary Sciences, viz. some of the wild animals sharing nearly identical infections of domestic livestock and disseminating the same to susceptible stock and mammals and birds kept in captivity in zoological gardens including edible freshwater and marine fishes, it has become difficult to compile all details of the parasitic problems of a large variety of hosts in one book. Moreover, with the intensification of husbandry practices separately for dairy animals (cows and buffaloes), sheep and goats, pigs, poultry and of others, it has become imperative to have host wise compilations.

Though buffalo is an important dairy animal in India and in many other countries, sufficient information is not available about its common parasitic diseases in textbooks on Veterinary Parasitology. The present book is, therefore, an attempt, without any claim of completeness, to highlight concisely important parasitic diseases of buffaloes reared in Indian subcontinent. It is assumed that students, teachers and clinicians dealing with this host species will be benefited to some extent in understanding various parasitic diseases of buffaloes, scope for further investigation on the unsolved problems as well as for adopting suitable control measures with the ultimate aim of raising production.

B.B. Bhatia

P.P.S. Chauhan

CONTENTS

LIST OF FIGURES

INTRODUCTION

Buffalo (*Bubalus bubalis*) is the most important dairy animal in India. Known to feature in the folklore, literature and religion of many lands, its domestication was in evidence some 4,500 years ago in the Indus valley civilization (Cockrill, 1970). The merits, versatility and contribution of buffalo to the agricultural economy have been stressed by a number of workers (Hafez, 1952; Shah, 1967; Cockrill, 1966, 1968, 1970, 1974; Sebastian *et al.*, 1970). Known to thrive and adapt remarkably under severest climatic conditions where other domestic stock could barely exist, buffalo excels cow in its ability to digest the coursest roughages, better utilization of Nitrogen and good Calcium and Phosphorus balances. Because of its immense value as a rich source of animal protein (milk, milk products and meat) and a potential draught animal, buffalo, in tropical countries, is receiving greater attention in dairy farming. In India, because of the ban placed in some states on cow slaughter, its meat, on account of lower rates, is extensively used by non-Hindu population.

Buffaloes are substantially distributed in more than 25 tropical countries—stretching from Southern Europe through India and China to the whole of South-East Asia. The world population of domestic buffaloes in 1979 has been estimated at 126 million of which 122 million are in Asia including 60.7 million heads of buffalo in India alone (Kim, 1981). Therefore, nearly half of the world buffalo population is in Indian sub-continent and yet scientific knowledge concerning this animal has not been commensurate with its increasing numbers and importance. According to Bhattacharjee (1964), the rate of annual increase in number of buffaloes, from 1956 to 1961, was 13.9% in comparison to 10.7% for cattle. India's contribution to world buffalo milk production, according to FAO production Year-Book (1971), was 59.05%. In Uttar Pradesh alone, more than 70% of milk was produced by buffaloes during 1977-78 (statistical data presented at Cattle Development Seminar, organised by Animal Husbandry Department, U.P. at Mathura, 1-3rd March, 1981). Buffaloes are, therefore, the main producers of milk and preferred species as indicated by the increasing ratio of she-buffaloes to cows in many parts of the country.

Food and Agricultural Organisation, emphasizing the importance of buffalo, expressed the need to stimulate interest in this species in India where, according to its Animal Health Year Book (1978), the total buffalo population was reaching sixty two millions. Cockrill (1966, 1970) rightly remarked that as a key animal for the hungry world and as a premier milk producer, buffalo alone ranks among the greatest sources of relief of India's food problems.

Buffalo diseases have been identified as one of the

major factors which have disrupted the development of the industry in Asia and have caused substantial economic loss to the poor subsistent farmers in the developing countries (Othman and Baker, 1981). Besides diseases and malnutrition, a major handicap in the effective development of buffalo husbandry appears to be the high figures of calf mortality—the highest rate being in calves below 6 months of age. The main causative factors, as reported by various workers, included pneumonia, salmonellosis and parasitism.

It is assumed that buffaloes and cattle have almost identical parasitic diseases. Much of the published information, therefore, is intermingled for both the species and consequently a detailed account of parasitic diseases, exclusively of buffaloes, is lacking. There are certain differences in the prevalence, clinical manifestations, pathological syndrome and other aspects of a number of parasitic infections in bovine and bubaline hosts. There are some parasitic infections which are encountered only in buffaloes e.g. "biliary emphistomosis" and "paracooperiosis" while to some others this host is either refractory or is acting as a reservoir. In many infections, the similarity of parasites for both cattle and buffalo is questionable and need experimental evidences including cross-transmission trials. Yet, the parasitic diseases are not less important in buffalo than other infectious diseases. These mainly include gastro-intestinal helminthoses, coccidiosis, fascioliosis and mange (Othman and Baker, 1981).

Among different bubaline parasitic infections, helminthic diseases are most varied and of common occurrence. Griffiths, in his chapter on "Parasites and Parasitic Diseases" in a FAO compilation on "The Husbandry and Health of Domestic Buffalo", edited by Cockrill (1974), stated that the buffalo is exposed to a higher risk of infection with snail borne helminths due to the animals propensity to seek rivers, pools or swamps for wallowing. Besides, other etological factors known commonly to affect the buffalo, diverse and manifold types of "helminthosis" and "helminthiasis", resulting from different grades of infections with flukes, tapeworms and roundworms, are responsible for marked deleterious effects that tend to lower the overall production both by way of morbidity and mortality. Certain protozoan infections are also emerging as important pathogens. External parasites also play an equally important role as in cattle.

Since the global review on the parasitic diseases of buffalo compiled by Mohan (1968), there has been a substantial addition in the literature on the parasites and parasitosis of Indian buffaloes. During the last two decades, much attention has been paid by different workers on major helminthic and protozoan problems to our dairy animals particularly of buffaloes. An attempt in the form of this compilation has been made to provide sufficiently the overall information available so far, on the important parasitic diseases of buffaloes and their calves in India, system-wise briefly. The relevant information particularly on the epidemiology, lesions, symptoms, diagnosis and treatment of important parasitic disease conditions have been highlighted. Such a compiled information would not only act as an essential pre-requisite for the formulation of suitable control measures with the ultimate aim of achieving increased production, but also provide clues to understand many unsolved or difficult problems of parasitic origin often faced by clinician/practising veterinarians and pathologists. It is hoped that the information would stimulate workers to conduct further studies towards a correct evaluation of important parasitic diseases of buffaloes in different geoclimatic zones of the country. The publications referred to in the text could be consulted by those interested in details.

References

1. Animal Health Year-Book 1978. Animal Health Service : Animal Production and Health Division. F.A.O., W.H.O. (ed. Konigshofer, H.O.).
2. Bhattacharjee, J.P. 1964. *Indian J. Agric. Eco.* **19** : 105.
3. Cockrill, W.R. 1966. *Ceylon Vet. J.* **14** : 49-51.
4. Cockrill, W.R. 1968. *Veterinarian. London.* **5** 265-272.
5. Cockrill, W.R. 1970. *Science J.* **6** : 35-40.
6. Cockrill, W.R. 1974. The *Husbandry* and *Health of Domestic Buffalo*, F.A.O. Pub., Rome 993 pp.
7. Hafeez, E.S.E. 1952. *Indian Vet. J.* **22** : 257-264.
8. Kim, Y.J. 1981. *Development of the Buffalo and Cattle Industry in South east Asia : An Economic Note in* Recent Advances in Buffalo Research and Development. FFTC Book Series No. 22. Taiwan, 1981, pp. 205.
9. Mohan, R.N. 1968. *Diseases* and *Parasites* of *Buffaloes.* Part IV. *Parasitic* and *Miscellaneous Vet. Bull.* **38** : 735-756.
10. Othman, A.M. and Baker, S.A. 1981. *Prevention and Control of Common Buffalo Diseases* : in Recent Advances in Buffalo Research and Development. FFTC Book Series No. 22. Taiwan, 1981.
11. Sebastian, L. Mudgal, V.D. and Nair, P.G. 1970. *J. Anim. Sci.* **30** : 253.
12. Shah, M.K. 1967. *Indian Fmg.* **17** : 39-42.

1

DIGESTIVE SYSTEM

1. Abomasitis (Gastritis)

Etiology

It is an important parasitic condition among ruminants including buffaloes caused by a number of trichostrongloid nematodes. Abomasitis can further be differentiated on the basis of the specific nematode infections as 'haemonchosis' 'Ostertagiosis' and 'trichostrongylosis'. The common species incriminated in abomasitis in buffaloes are : *Haemonchus contortus, H. bubalis, H. placei, Mecistocirrus digitatus, Ostertagia ostertagi, Skrjabinagia boevi* and *Trichostrongylus exei* (Sharma and Pande, 1963; Chauhan and Pande, 1968; Pande and Chauhan, 1969).

Epidemiology

As a whole, the incidence of different gastrointestinal strongyloid (bursate) nematodes remains maximum during rainy and early winter seasons (Fig. 1). Buffalo-calves, between 6-9 months age, carry the maximum worm-load (Fig. 2) and a little less in calves between 3-6 months age. The rate of incidence reduces gradually in calves between 9-12 months age and above (Chauhan, Bhatia and Pande, 1973). The animals become infected with these nematode parasites by ingesting their larvae during grazing. The development of these larvae and their survival on the ground depends on a number of extrinsic factors like temperature and humidity, soil, vegetation etc. Optimum conditions for the development of pre-parasitic larval stages of these nematodes differ from species to species but moisture in the soil is more important. In nature, mixed infections are the rule. Usually every animal 's parasitised by a mixture of a number of species in repeated doses. Specific disease identities are as under.

(*i*) *Haemonchosis*

It is caused by several species of *Haemonchus* viz., *H. contortus*, *H. bubalis* and *H. placei*. These are voracious blood suckers and their main deleterious effects on the animal are due to loss of blood. Both parasitic larvae or juveniles and adults suck heavy quantity of blood. In heavy infections, the gastric mucosa is severely damaged and symptoms like anaemia, emaciation, oedema, marked swelling under the jaws, besides digestive disturbance, are the clinical manifestations of the disease. The gastric mucosa is seen swollen with peticheal haemorrhages and shallow ulcers, on autopsy. Besides *H. contortus* and other species,

H. bubalis is more frequently seen in bunches in heavy infections of abomasum in our buffaloes. Its incidence remains maximum during rainy season followed by winter and summer. The worms remain partly embedded into the mucosa which is seen congested, inflamed and oedematous covered with thick mucoid exudate.

Likewise, *Mecistocirrus digitatus* can also become one of the main causes of severe gastritis in calves.

(*ii*) *Ostertgiosis*

It is caused by *Ostertagia ostertagi* and *Skrjabinagia boevi*. *O. ostertagi* is a brown stomach worm and is another blood sucker. The worms form small, circular raised areas of about 1-2 mm in diameter on the mucosa. In these nodular lesions, the worms develop and the adults may be seen partly projecting through a small opening in the centre or may also be found free on the mucosa. It is mostly seen in cattle than in buffaloes.

Skrjabinagia boevi is more frequently encountered in buffaloes than *O. ostertagi* in the abomasa of buffaloes. Surveys revealed that *S. boevi* infection was found in nearly half of the buffaloes examined. The maximum incidence was observed to be during months of January and February (Pande and Chauhan, 1969). In *S. boevi* infection, the fundic and pyloric regions of the abomasal mucosa show characteristic lesions from invasion by the adult worms and/or their earlier developmental stages. Gross lesions include slightly raised whitish circular areas of 1-2 mm diameter with depressed centre clearly suggestive of nodulation which remain covered with thin layer of mucus. In the pyloric regions, the gross lesions consist of slightly oedematous and eroded patches covered with thick, pale translucent mucus. Histological studies of such lesions depict all inflammatory and tissue reactions of acute gastritis. The juveniles remain deep inside the gastric glands causing severe damage. Marked infiltration with leucocytes (particularly lymphocytes and oesinophits) around affected glands and congestion of sub-mucosal blood capillaries are generally seen (Fig. 3).

(iii) *Trichostrongylosis*

It is caused by *Trichostrongylus axei* which is a highly pathogenic nematode when present in large number. The minute worms usually do not suck blood and remain mostly inside the characteristic circumscribed slightly hard and raised areas of about 1.5 cm in diameter covered with mucous in the fundic region. The affected area of the abomasum may be inflamed and corrugated. There may be sloughing of epithelium and also whitish necrotic debris on the surface.

The young calves are much more susceptible to these infections than older animals, but the latter become the source of the infection for the former. Calves grazing alongwith adult animals may be heavily infected and suffer from serious illness. Malnutrition and debilitated condition may predispose the animals for acquiring clinical conditions.

Diagnosis

Diagnosis of the above disease conditions can be made by finding eggs through faecal examination, faeces culture for larvae, lesions detected during postmortem cases and recovery of worm.

2. Enteritis

Etiology

Different regions of small and large intestine are parasitised by a variety of nematodes, cestodes, blood flukes, immature stages of amphistomes and coccidia, singly or in mixed infections, resulting in complicated disease condition—the "parasitic enteritis". This can be differentiated as 'immature amphistomosis', 'neoascariosis', 'bunostomosis', 'paracooperiosis', 'visceral schistosomiasis', 'oesophagostomiosis', and other among helminthic diseases and 'coccidiosis'. The intensity of one or more of these parasitic conditions of the intestine depends on a number of factors viz. age of the animals, their general physical condition, level of nutrition, dose of infection, possibilities of repeated exposure to infection, multiplicity of infections, climatic condition to which animals are exposed, management and hygiene etc. These different types of parasitic enteritis in buffaloes are of common occurrence in India and are briefly descibed below :

(*i*) Immature amphistomosis

Epidemiology

This disease is widely prevalent in the country and is caused by immature stages of more than a dozen species of amphistomes (Patnaik, 1964; Pande, 1935; Malik and Sharma, 1964; Gupta, *et al*, 1978). Just after excystation of a large number amphistome metacercariae ingested by the susceptible host, the immature stages immediately attach firmly to the mucosa of upper portion of small intestine on way to rumen to bear the against stream of ingesta. The disease is commonly seen among young calves left for grazing near water collections infested with snails just after rains. With the intake of large number of metacercariae, the disease comes up as an outbreak, in endemic areas, during winter season. Due to mass attachment of immature flukes by their powerful suckers, the mucosa of the duodenum is acutely inflamed and hypertropiced (Fig. 4). Haemorrhage occurs, due to disruption of submucosal blood capillaries. Symptoms like dark-coloured foetid diarrhoea, oedema under the jaws, anaemia appear suddenly in young calves which collapse in large number in severe infection mainly due to dehydration. Presence of gastro-intestinal nematodes and coccidia, invariably seen in outbreaks of amphistomosis, become additive factors for bringing fatal condition.

Of a variety of amphistomes, *Paramphistomum* (*Explanatum*) *explanatum* is most commonly involved, besides causing enteritis, in parasitizing biliary passages and liver of buffaloes. Specially in buffaloes, this species of amphistome reaches the bile duct, gall bladder and liver. Kulasiri and Seneviratna (1956) believed that the infection of the bile duct was by the transperitoneal route. Pande, Bhatia, and Arora (1968), recorded the juveniles of this amphistome from pancreatic duct also in addition to their location in enlarged and thickened bile duct opening into the small intestine. According to them, the normal route of enterance appeared to be through the openings of the ducts while, on reaching deep into the submucosa of duodenum, some immature specimens could, prior to their nodulation, be transported to other organs through portal system. Describing the pathology, the authors detected blockade of bile and pancreatic ducts with mucosal plugs inside the acetabulum glands and even necrosis at places. The contents of the intestinal caeca and pharynx suggested that these flukes feed mostly on the mucosal debris and blood.

In adult buffaloes, though such severe lesions are usually not observed, enteritis is evident due to mass attachment by the immature flukes of which some may be seen quite deep in the mucosa reaching sub-mucosa or muscular layers where most of these are entrapped and leave greenish nodules in the small intestine. Such nodules in the region of small intestine should be differentiated with nodules formed by the fourth stage larvae of *Oesophagostomum radiatum* causing 'pimply gut'.

(*ii*) Neoascariosis

Etiology

It is caused by the large intestinal worm, *Neoascaris vitulorum* Among round worm infections, 'neoascariasis', is widely prevalent disease of buffaloes and cattle in India. It is one of the greatest menace in rearing healthy calves. Because of its unique pattern of life cycle, the mode of infection being prenatal and or transmum the newly born calves suffer most due to this infection. Mostly, 'neoascariosis' becomes a single major factor for heavy mortality among calves.

Incidence

There are number of reports on the incidence of *N. vitulorum* infection among cattle and buffalo calves both from our country and abroad. The highest incidence of *N. vitulorum* is commonly seen during summer months and among the youngest age group (upto 3 months) of calves, comparatively more in buffalo calves than in cattle calves. Severe ascariosis among calves has been a major parasitic problem in India (Das and Singh, 1955). Dass *et al* (1961) believed that ascariosis was a very serious problem in calves especially those of buffaloes and the young stock succumbed within 2-3 months after birth. Vaidyanathan (1949) found ascarid infection in 31-40 days old buffalo and cattle calves at Coimbatore, the youngest of the infected calves being 24 days old. A maximum of 400 worms were collected from a buffalo calf below 3 months by Srivastava (1963) who also encountered another case of intestinal perforation with location of worms in liver in three calves. A high incidence of *N. vitulorum* in one month old buffalo-calves was reported by Patnaik and Pande (1963).

Mode of Infection

A second larva within egg is the infective stage of *N. vitulorum* developed in the open under favourable conditions of temperature and moisture. Apparently, the animals become infected by ingesting these infective eggs through feed or water either in the shed or on grazing areas. There is no possibility of the worms becoming mature in adult animals or calves as the attempts to produce patent infections in these animals have failed. However, patent infection can be established by feeding infective eggs within twentyfour hours after birth. Possibly, the development cycle of *N. vitulorum* inside the host may be similar to that of *Toxocara canis* in dogs where the larvae pass from the mother to the foetus *via* the placenta. Besides, recovery of *N. vitulorum* larvae in milk of buffaloes (Tongson, 1971; Chauhan, *et al* 1974; Gautam *et al.*, 1976) and studies conducted by Mia *et al* (1975), proves that the route of infection of buffaloe-calves is also transmammary via the colostrum where the dam effectively serving as an intermediate host for the parasite. Larvae of *N. vitulorum* could also undergo partial migration from intestinal lumen to liver and lungs producing extensive damage as studied in albino mice, albino rat, guinea pig and chicken (Chauhan and Pande, 1972 and Chauhan *et al*, 1973).

Symptoms

N. vitulorum infection is usually a disease of young calves below 3 months of age. In heavy infections, the calves become extremely emaciated, unthrifty, unable to stand and passing foul-smelling clay-coloured watery faeces. Prior to it, there may be symptoms of verminous pneumonia due to larval migration through lungs. On autopsy of clinical cases, a large number of worms with light yellowish and sticky intestinal contents and extensive erosion on the lining resulting from mechanical damage are invariably noticed.

Diagnosis

By finding eggs in the faeces, symptoms and worms recovered in faeces or in postmortem cases.

(*iii*) **Strongyloidiosis**

Etiology

It is caused by *Strongyloides papillosus*, the common intestinal thread worm of domestic ruminants, which is widely prevalent and responsible for severe enteritis in heavy infection.

Epidemiology

As much as 56.79% of buffalo calves upto 1 year have been found to harbour this infection among calves from 3-6 months age (Chauhan *et al*, 1973). Patnaik and Pande (1963) also ranked *S. papillosus* second in frequency and pathogenicity to *N. vitulorum* in young buffalo-calves under 5 weeks of age. The parthenogenetic female penetrates deep in the mucosa of small intestine (particularly in region of duodenum and jejunum) and lay strings of eggs inside it (Fig. 5). In heavy parasitization, there is catarrhal inflammation with petechial and ecchymotic haemorrhages, erosion and necrosis of mucosa with extensive muscus production in duodenum. Infection usually takes place within the sheds as the infective larvae infect the susceptible hosts through active penetration particularly through the skin of foot. Intermittent diarrhoea with mucus and blood is a common symptom. There is loss of condition, retarded growth and sometime death of the calf in heavy infection. The infective larvae may cause dermatitis when these penetrate the skin and secondary infection may result. In nature, *S. papillosus* infection is found invariably mixed with other intestinal helminths and coccidia.

Diagnosis

By finding eggs in the faeces and worms on postmortem examination.

(*iv*) **Bunostomiosis**

Etiology

Indian buffalo has been reported as a host of 3 hookworms, *Bunostomum phlebotomum.*, *B. trigonocephalum* and *Gaigaria pachyscelis* responsible for bunostomiasis. Sharma and Pande (1963) studied hookworm infection in adult buffaloes while Chauhan (1972) reported the common occurrence of *B. phlebotomum* in buffalo-calves (38.12%).

Lesions

Grossly, the worms remained attached deeply to the mucosa of the small intestine with their powerful buccal capsule, lacerate the mucosa and cause it to bleed. There is thickening and oedema with production of thick mucus. Freshly detached feeding spots reveal erosion and haemorrhage. There may be large number of haemorrhagic nodules, in heavy infection, of pin-head to millet seed-sized with ecchymotic swellings projecting towards the intestinal lumen. On teasing, the nodules yield fourth stage juveniles of *B. phlepotomum*. The intestinal contents may be seen mixed with blood.

In young animals, infection with large number of hookworms causes severe haemorrhagic enteritis which may be aggravated with other concurrent helminthic infections of the regions of the small intestine. The animals show diarrhoea, emaciation, severe anaemia, loss of weight, and oedema of intermandibular space. Severe signs of hookworm infection in young animals appear before the worms inside the intestinal lumen reach the maturity and their eggs are passed out.

Diagnosis

By finding egg in faeces and their culture. Recovery of worms on post mortem examination.

(*v*) Paracooperosis

Etiology

Paracooperiosis is another widely prevalent nematode infection of intestine of buffaloes in the country. Its causative agent, *Paracooperia nodulosa*, is very pathogenic during its histotropic phase resulting in extensive nodulation of a large part of the small intestine. Among many workers in India, Nair and Anantaraman (1954), Sharma and Pande (1963), Patnaik and Pande (1963), Venkatratnam (1964), Sundram *et al* (1968), Khadse and Dubey (1968), Achutan (1970), and Chauhan and Pande (1972a) have described extensive nodulation of the small intestine varying in size from a poppyseed to the hazalnut but commonly of pea size with symptoms of dullness, emaciation, anaemia, diarrhoea, oedema of the lower jaw culminating in death and the histopathology of the lesions.

Epidemiology

The data given by Chauhan and Pande (1972b), on the seasonal incidence of *P. nodulosa* in buffalo calves upto 1 year of age revealed that the presence of nodules was maximum (92.85%) in the summer months and minimum (46.05%) during the winter season (Fig. 6); the presence of adult worms in the lumen was maximum in winter season (57.14%) and minimum during summer season (15.38%) while nodules were found in all cases (100%) during summer followed by rainy and winter season (85.36% and 60% respectively) (Fig. 7), among positive cases. The nodules though commonly present, showed variation in their number, size and distribution throughout the year. The number in a single case, varied from 8-1063. In size, these nodules could be distinguished into 3 categories—pin-head, millet and peaseed sized. The first appearance of the small-sized nodules encountered usually during early summer to the early rainy season. The size and number of nodules continued to increase in subsequent months. In late winter, adult worms had emerged from most of the nodules. These observations, therefore, clearly indicate that susceptible hosts mostly pick up the infections during the late winter and spring seasons when there is massive development of free-living stages which ultimately give rise to extensive nodulation during summer.

Lesions

Grossly, the lining of the infected intestines exhibit copious, thick mucofibrinous exudate particularly when the adult worms have penetrated into the mucosa. The nodules of different sizes are easy to detect as raised rounded bodies and are also visible from serosal side. On teasing large nodules, the central part exhibits dark haemorrhagic spots while others have an opening indicative of the emergence of larvae. Much older and fibrosed nodules with opening towards the lumen show inspissated yellowish pus with some calcification. The smaller nodu leslook reddish white to light yellow in colour. On histological examination, the youngest nodules reveal the juvenile surrounded by thick zone of inflammatory cells and damage of mucosal and submucosal layers with a tract. The medium sized nodules also involve muscular layer with a 4th stage in a small cavity surrounded mainly by lymphocytes and eosinophils alongwith red blood cells (Fig. 8). There is beginning of proliferation of fibroblests. Older nodules demonstrate extensive disruption of all the layers of the intestine and fibrosis with fewer infiltrating cells in the adjacent areas.

Considering the pathogenic significance of *P. nodulosa* among growing calves, the problems need further consideration on lines of studies conducted in *Oesophostomum columbisnum* infection in sheep where there is reduction of absorptive surface by extensive nodulation, retardation of paristaltic movements of intestine absorption of toxic endproducts of nematodes resulting in partial paralysis of the intestine which ultimately lead to poor growth of the animal. The clinical picture in paracooperiosis is yet to be studied under controlled conditions and pathognomic symptoms established.

Diagnosis

As in bunostomiasis.

(*vi*) Visceral schistosomiosis

Etiology

In buffaloes, it is generally considered that blood flukes are usually unaccompanied by any marked pathological changes and probably this host species is more suitable definitive host than cattle and other hosts. Commonly, *Schistosoma spindale* and occasionally *S. indicum* are involved in visceral schistosomiosis in buffaloes. The flukes occur in the mesentric and portal blood vessels and may also reach liver in heavy infection. Eggs of *S. spindale* are mainly responsible for enteritis in calves.

Lesions

In adult buffalo, Sharma and Pande (1963) and Ahluwalia, Pande and Bhatia (1966) described the histopathology of grossly looking blackish pseudoabscesses in the caecal mucosa. Characteristic eggs can easily be detected in the scrapings from haemorrhagic spots of varying size distributed irregularly on the mucosal lining of both small and large intestines. The pathological changes, around eggs in such lesions in adult buffalo, are comparatively less pronounced. In heavy infection, the invasion of liver by blood-flukes may also occur.

The pathological picture in buffalo calves is, on the other hand, quite different. Pande, Bhatia and Arora (1968) clearly demonstrated the presence of numerous pseudotubercles both in the small intestine and regions of the large intestine, the former being more severely affected. Histologically, the lesions revealed marked tissue reactions around the entrapped eggs (Fig. 9). In slightly older tubercles, there was definite zone of fibrosis. In much older tubercles, the egg-shell alone was left over from phagocytosis due to the activities of numerous macrophages. The common infiltrating cells encountered were eosinophils, lymphocytes, giant cells and some macrophages. Besides, congestion of blood vessels in the submucosa and erosion of the surface epithelium of villi with their destruction were among the other pathological changes. From these observations, it is proved that blood flukes evoke marked pathogenic effects in buffalo also particularly in younger age group.

Orientobilharzia dattai is another blood fluke reported from buffaloes (Dutt and Srivastava, 1952, 1961 and 1962). Pathogenic effects are less severe particularly in adult buffaloes.

Diagnosis

By finding eggs in the faeces. Lesions and recovery of worms in postmortem cases.

(*vii*) Oesophagostomiosis

Etiology

Oesophagostomum radiatum, the nodular worm of cattle, is another important pathogenic nematode responsible for extensive nodulation of intestinal tract particularly of the large intestine of buffalo. It is cosmopolitan in distribution. As high as 71% of incidence of this nematode in buffalo during summer months has been observed (Chauhan, 1972).

Epidemiology

O. radiatum infection occurs by ingestion of infective ensheathed larvae during grazing on contaminated pastures. Calves, exposed soon after rains, acquire a heavy worm burden. The principal effects occur during the prepatent period and are caused by the larvae. Nodulation of the intestine is a severe tissue reaction that follows repeated infection. Following initial infection, small raised areas of about 1 mm diameter in the intestinal wall are formed due to penetration of parasitic larval stages. They grow into small abscesses filled with leucocytes and becomes double the size. The intestinal wall also becomes inflamed and oedematous. Following repeated infections, the nodules grow larger reaching upto 5 mm in diameter containing leucocytes, giant cells surrounded by fibroblasts.

In clinical condition, the affected animals show diarrhoea followed by anoraxia and tenesmus, rise in temperature, loss of weight, emaciation and marked decrease in erythrocytes probably due to toxin produced by the worms. The main pathogenic effects are caused by the extensive nodulation of a larger area of large intestine, particularly caecum and colon. In massive infection, small intestine is also involved. Thus, principal effects occur during prepatent period caused by the larvae. Seasonal incidence of this worm can be substantially reduced if the positive cases are made free of this infection before onset of rains to effective chemotherapy

A much detailed account is available on the pathogenesis of *O. radiatum* in cattle but not in buffaloes.

Diagnosis

By faecal examination for the detection of eggs, by faeces culture and by detection of typical nodular lesions and recovery of worms or their juveniles at autopsy or postmortem examination. In culture, the third stage larvae of *O. radiatum* can easily be differentiated from those of other gastro-intestinal nematodes by features like longest sheath (old cuticle), maximum extension of sheath beyond tail and 20 intestinal cells beside other characteristics (Chauhan, Bhatia and Pande, 1973).

(*viii*) **Other helminths**

There are many other helminths responsible for enteritis in buffaloes and their calves viz. *Trichostrongylus* spp. *Cooperia* spp. and *Trichuris* spp. (*T. discolor* recorded in buffaloes in M.P. by Bhopale *et al*, 1971 a), among nematodes and *Moniezia expansa*, *M. benedeni*, *Stilesia* sp. and *Avitellina* sp. (*A gaughi* was recorded in buffaloes in M.P. by Bhopale, *et al* in 1971), among cestodes. They also produce many digestive disturbances and lesions in the intestine by penetrating into the mucosa in heavy infections. As regards cestode infections, light infections are of less significance. Only in massive infection, ill effects are seen in younger animals. Infection takes place by ingestion of oribatid grass mites which act as an intermediate hosts for these anoplocephalid tapeworms of ruminants. Generally, buffaloes are relatively infrequent hosts of tapeworms (Mohan, 1968). Calves, heavily infected with tapeworms, may become weaker, their coats become rough and their growth may be retarded. There may also be oedema of the abdominal region giving potbellied appearance and anaemia may follow. There is alternate diarrhoea with constipation. Bunches of tapeworms may also occlude the intestinal lumen.

Diagnosis

By finding eggs in the faeces, lesions and recovery of worms at autopsy or post-mortem examination.

Treatment

In general, gastro-intestinal roundworms in all types of domestic ruminants continue to infect them in spite of a plethora of safe and efficacious anthelmintic drugs. A thorough information of the biology of the worms and of the climatic conditions of the area is necessary before the use of anthelmintic drugs. One may treat the entire herd at regular 'strategic' times in order to prevent the build-up of infective forms in the pastures and, therefore, prevcnt the outbreaks of clinical gastro-intestinal helminthoses.

Valid information on treatment of various gastro-intestinal parasites in buffaloes is not available. However, a number of anthelmintics which are recommended for treating gastro-intestinal nematodiosis, and immature paramphistomosis in domestic ruminants are appended below :

(A) **Gastro-intestinal Nematodiosis**

1. *Phenothizine* (*Phenovis ICI*) : Effective against a number of roundworms particularly causing abomasitis.
2. *Thiabendazole* (*Thiabendole MSD*) : Effective against both immature and mature nematodes causing abomasitis and enteritis especially *Hae-*

monchus, *Ostertagia*, *Trichostrongylus*, *Bunostomum*, *Strongyloides*, *Oesophagostomum*, *Trichuris*.

3. *Neguvon* (*Bayer L 13/59*) : Effective against certain nematodes especially *Haemonchus*, *Ostergagia*, *Trichostrongylus exei* and *Oesophagostomum*.
4. *Parbendazole* (Helatac SKF) :
5. *Tetramisole* (*Nilverm ICI*) : Effective against *Haemonchus*, *Ostertagia*, *Trichostrongylus*, *Bunostomum*, *Oesophagostomum* as well as lungworms.
6. *Pyrantel tartrate* (Binminth I, Pfizer).
7. *Morantel tartrate* (Binminth II, Pfizer).
8. *Fen-bendazole*.
9. *Cam-bendazole*.
10. *Bephenium embonate* and *Bephenium hydroxynaphthoate*.
11. *Methyridine* (Promintic) : Effective against both immature and mature worms particularly *Trichostrongylus*, *Strongyloides*, *Ostertagia* and *Haemonchus*.
12. *Levamisole*
13. *Piperazine compounds* (Piperazine citrate, P. adipate, P. dihydrochloride etc.) effective against *Neoascaris vitulorum*.
14. Haloxon etc.

(B) Immature Paramphistomosis

In buffaloes, clinical trials against immature amphistomosis has been done by some workers viz. Chhabra and Gill (1975), Chhabra and Bali (1976) and Panda and Misra (1980) with few of the anthelminthics with variable efficacy.

1. *Hexachlorophene*.
2. *Niclosamide*

3. *Hexachloroethane*
4. *Oxyclozanide*
5. *Manichlopholan.*

C. Visceral Schistosomiosis

There is as yet no report on the treatment of visceral schistosomiosis in buffaloes. However, antimonial drugs, used in nasal schistosomiosis and in schistosome infection of other hosts, may be useful.

References

1. Achutan, H.N. 1970. *Indian Vet. J.* **47** : 451-53.
2. Ahluwalia, S.S.; Pande, S.P. and Bhatia, B.B. 1966. *Some of the Chief Pathogenic Helminths of Adult Buffalo in Uttar Pradesh, Deptt. Bull. No. II. Coll. Vet. Sci. & A.H.*, Mathura, 3 pp.
3. Bhopale, K.K.; Joshi, S.C. and Kamalapur, S.K. 1971. *Indian Vet. J.* **48** : 1282-83.
4. Bhopale, K.K.; Joshi, S.C. and Kamalapur, S.K. 1971a. *Orissa Vet. J.* **6** : 104-106.
5. Chhabra, R.C. and Bali. H.S. 1976. *J. Res. PAU.* **13** : 226.
6. Chhabra, R.C. and Gill, B.S. 1975. *J. Res. PAU.* **12** : 184-188.
7. Chauhan, P.P.S. and Pande, B.P. 1968. *Ceylon, Vet. J.* **16** : 74-76.
8. Chauhan, P.P.S. 1972. Ph.D. Thesis. Agra Univ., Agra.
9. Chauhan, P.P.S. and Pande, B.P. 1972. *Indian J. Exptl. Biol.* **10** : 193-201.
10. Chauhan, P.P.S. and Pande, B.B. 1972a. *Indian J. Anim. Sci.* **42** : 930-34.
11. Chauhan, P.P.S. and Pande, B.P. 1972b. *Indian J. Anim. Sci.* **42** : 1033-38.

12. Chauhan, P.P.S.; Bhatia, B.B. and Arora, G.S. and Agarwal, R.D. 1973. *Indian J. Anim. Sci.*
13. Chauhan, P.P.S.; Bhatia, B.B. and Pande, B.P. 1973. *Indian Anim. Sci.* **43** : 216-19.
14. Chauhan, P.P.S., Agarwal., R.D. and Ahluwalia, S.S. 1974. *Curr. Sci.* **43** : 486-87.
15. Das, K.M. and Singh, G.B. 1955. *Brit. Vet. J.* III: 342-47.
16. Dass, N., Panda, S.N. and Biswal, G. 1961. *Indian Vet. J.* **38** : 194-97.
17. Dutt, S.C. and Srivastava, H.D. 1952. *Parasitol.* **42** : 144-149.
18. Dutt, S.C. and Srivastava, H.D. 1961. *Indian J. Vet. Sci.* **31** : 288-303.
19. Dutt, S.C. and Srivastava, H.D. 1982. *Indian J. Vet. Sci.* **31** : 288-303.
20. Gupta, P.P., Singh, E. and Dutt, S.C. 1978. *Indian Vet. J.* **55** : 491-492.
21. Gautam, O.P., Malik, P.D. and Singh, D.K. 1976. *Curr. Sci.* **45** : 350.
22. Khadse, K.T. and Dubey, J.P. 1968. *JNKVV Res. J.* **2** : 63-64.
23. Kulasiri, C. and Seneviratna, R.D. 1956. *J. Comp. Path.* **66** : 83-88.
24. Malik, P.D. and Sharma, R.S. 1964. *11th Conf. Anim. Dis.* Madras.
25. Mia, S., Dewan, M.L., Uddin, M., Chowdhury, M.V.A. 1975. *Trop. Anim. Hlth. Prod.* **7** : 153-56.
26. Mohan, R.N. 1968. *Diseases* and *Parasites* of *Buffaloes. Part. III. Parasitic and Miscellaneous Diseases Vet. Bull.* **38** : 735-56.
27. Nair, K.P.C. and Anantraman, M. 1954. *Indian Vet. J.* **30** : 334-35.
28. Panda, B.K. and Misra, S.C. 1980. *Indian J. Anim. Hlth.* **19** : 131-35.

29. Pande, B.P., Bhatia, B.B. and Arora, G.S. 1968. *Indian J. Vet. Sci.* **38** : 463-70.
30. Pande, B.P. and Chauhan, P.P.S. 1969. *Indian J. Anim. Sci.* **39** : 79-88.
31. Pande, P.G. 1935. *Indian J. Vet. Sci.* **5** : 364-75.
32. Patnaik, M.M. 1964. *Indian J. Helm.* **16** : 48-51.
33. Patnaik, M.M. and Pande, B.P. 1963. *Indian Vet. J.* **40** : 128-33.
34. Sharma, K.M.L. and Pande, B.P. 1963. *Indian J. Vet. Sci.* **33** : 42-50.
35. Srivastava, S.C. 1963. *Indian Vet. J.* **40** : 758-62.
36. Sundaram, R.K. Sivasubramaniyam, M.S., Iyer, R.P. and Peter, C.T. 1968. *Indian Vet. J.* **45** : 22-31.
37. Tongson, M.S. 1971. *Philippine J. Vet. Med.* **10** : 60-63.
38. Vaidyanathan, S.N. 1949. *Indian Vet. J.* **26** : 228-30.
39. Venkataratnam, A., Padmavathi, P. and Indra Devi, T. 1969. *Indian Vet. J.* **46** : 261-62.

(*ix*) **Coccidiosis**

Etiology

The disease, coccidiosis, occurs in many domestic and wild animals and is responsible for a variety of enteritis in their hosts. The agents which cause coccidiosis are protozoa (tiny one-celled animals) chiefly of the genus *Eimeria.*

Coccidia constitute an important group of sporozoan parasites associated with considerable morbidity and, in severe cases, causing mortality. Undoubtedly, it is one of the major problems facing the animals industry particularly in view of the intestification programmes undertaken. As intracellular parasites in the lining of the digestive tract and the associated structures, their immense reproductive potentialities often occasion varying grades of disorders specially because of their tissue destroying habits. The extent of the injurious effects is naturally dependent on the differential details in pattern of their life cycles. Because of the direct life-cycle which involves alternation of asexual and sexual generations inside the host, coccidial infections are easily acquired as a result of overcrowding—a natural outcome of confining the animals during domestication in limited areas. Insanitory and unhygienic conditions, coupled with backward husbandry practices make the animals easy victims to 'Coccidiosis'. It is the younger stock that usually suffer most from these infections. Older lot, seldom manifesting clinical symptoms, acts as carrier and, in chronic cases, becomes emaciated which lowers the overall production.

Pathogenic Significance

The loss in body weight, reduction in milk yield, production of inferior wool fibres, toxic role of coccidia, relationship of the disease with paralysis and nervous symptoms have all been known from the work of scientists (Malikyan, 1955; Lotze, 1959; Shumard, 1957; Caranti, 1956; Guillon, Palisse, and Toucas, 1959). Coccidian symptoms become more pronounced when concurrent infections of the digestive tract with certain intestinal nematodes are also present which commonly occur. It is now being realized that ihe disease entity has become a major factor responsible for above mentioned conditions in lambs, kids and cattle and buffalo-calves. Diarrhoea, with or without haemorrhage, if accompanied by occystic discharge, constitute the usual criteria in diagnosing the field cases. This diagnostic method has several pitfalls. The age of the attached animals, its plane of nutrition, the infective occystic dose, the nature of the asexual and sexual development and their location associated with various species, and the concurrent infections harboured, produce varying degrees of pathological conditions. At times, the attendent clinical picture might fail to yield, during faecal examinations, the oocysts. A clinician, charged with treatment and control, needs be fully conversant with the pathogenic species and the nature of their development. Furthermore, in detecting and diagnosing the cases suspected of "entero-toxaemia" which fail to demonstrate the organism, coccidiosis must invariably be first elimated. Additionally, the two conditions viz. 'Coccidiosis' a clinical condition with typical manifestations of the disease and 'coccidiasis' a positive case carrying the infection without clinical manifestations should be differentiated.

General Pattern of Morphology and Life-cycle

In domestic ruminants, the coccidian species, based on their oocystic morphology, belong to the genus *Eimeria*. As a group, coccodia are generally considered with strict host or even organ specific. Identical oocysts are, however, encountered in the dropping from allied hosts eg. the coccidia of cattle and buffalo, sheep and goats and domestic fowl and some of the gallinaceous birds. Cross transmission and immunologicalst udies, so far conducted have been found with variable results.

The general pattern of morphology of oocysts of eimerian species is the same in all species. Differentiation of specific species is based on variations in shape, size, colour, presence or absence of certain structures and also on the variations of their endogenous stages inside their hosts.

Infection of the host occurs by ingestion of numerous

sporulated occysts through water and feeds. On reaching the alimentary tract, the oocysts are acted upon by the digestive juices resulting in the release of the sprozoites. Schizogony (asexual cycle) commences after the entry of the sprozoites in the host cells which may be either epithelial cells of villi and/ or crypts or endothelial cells lining the lacteals inside the core of the villus or in the connective tissue cells. Inside such cells, the sprozoites multiply by repeated binary fissions and, in a short period, develop to mature schizonts which are enclosed by a wall and contain 10 to many hundreds or thousands or hundreds of thousands of nearly benana-shaped merozoites. Resulting mature schizonts are of different varieties, mainly of two types (*i*) giant schizonts which may be visible with naked eyes containing millions of merozoites and (*ii*) epithelial schizonts which may be large containing 50 to few hundreds of merozoites or small containing few merozoites. These merozoites are released after bursting of the schizonts and attack fresh host-cells. There may be subsequent generations of schizonts but mostly epithelial types with different numbers of merozoites. In *Eimeria bovis*, in cattle and buffalo, there are two generations of schizonts—the first producing more than 100,000 merozoites and the second 30-36 merozoites. Thus, an infestive oocyst (containing 8 infective sporozoites), when ingested by a calf, has the potentiality of yielding aproximately 24 million second generation merozoites ($8 \times 100{,}000 \times 30$). If this potentiality is fully realised, ingestion of even 1,000 such oocysts could result in the destruction of 24 million intestinal cells (Hammond, 1964).

Following schizony, the gametogony (sexual cycle) starts when the merozoites develop to become male and female gamonts which ultimately result in the formation of oocysts. Such oocysts are detached from the lining and pass outside the host which, in open, undergo sporogony to become sporulated which is the infective stage. The complete cycle inside the host differs from species to species and may take from few days to few weeks.

Epidemiology

Coccisiosis in cattle has long been held responsible for exceedingly heavy economic loss in many advanced countries. The disease may be acute or chronic in nature occurring in both grazing and stall-fed herds (Pellerdy, 1974). It is one of the significant causes of mortality and retardation in growth in zebu cattle in the United States (Hammond, 1964). From India, bovine coccidiosis has been reported from time to time in our cattle (Couper, 1927: Naik, 1927; Rao and Hiregaudar, 1953-54) in outbreaks causing red dysentry in dairy cattle. The first report on coccidiosis in buffalo was of Biswal (1948) who ranked coccidiosis among epizootics of buffalo herds. Since last 15 years, sufficient work on different aspects on coccidia and coccidiosis of buffalo was conducted in the country (Patnaik and Pande, 1965; Bhatia and Pande, 1967; Pande, Bhatia, Chauhan and Garg, 1968; Pande, Bhatia and Chauhan, 1971; Bhatia, Pande, Chauhan and Garg, 1971; Shastri, Krishnamurthy and Ghaffor, 1974 and 1979; Shastri and Krishnamurthy, 1975). The disease is primarily of calves particularly below one year of age. Adult buffalo and cattle usually act as symptomless carriers and continuously pass out oocysts. In all, 13 eimerian species are known from buffaloes in our country based on their morphology (Bhatia, Pande, Chauhan, and Arora, 1968; Mandal, 1980).

1. *Eimeria alabamensis*
2. *Eimeria auburnesis*
3. *Eimeria bareillyi*
4. *Eimeria bovis*

5. *Eimeria brasiliensis*
6. *Eimeria bukidnonensis*
7. *Eimeria canadensis*
8. *Eimeria cylindrica*
9. *Eimeria ellipsoidolis*
10. *Eimeria pellita*
11. *Eimeria subspherica*
12. *Eimeria wyomingensis*
13. *Eimeria zuernii*

Symptoms

In acute cases, symptoms are mostly associated with severe foul smelling reddish dirrhoea (sanguinolent faecal discharge) due to haemorrhage in the intestine. The severity of symptoms depends on various factors viz. the age of the animal, the dose of infection in addition to a number of intrinsic and extrinsic conditions of the animal. In some, the stools may be just soft, others may pass watery stools with or without blood clots. The animal, at one time, may show recovery but there is sudden flare-up of the symptoms associated with recurrent diarrhoea, anorexia, tenasmus, inability to rise on legs, debility with or without fever. Young calves, in acute condition, die within a few days particularly when the stools carry large quantities of blood. This condition in young (suckling) calves needs to be differentiated from bovine dysentry and in older calves and adults from anthrax.

Bovine coccidiosis can be distinguished into three periods : (*i*) first 5-7 days showing debility, diarrhoea and loss of body weight, substantial reduction of milk yield or even no yield and clots of blood with mucus shreds present in the loose faeces with partial paralysis of the anal sphinctor resulting in its mostly wide open condition and (*iii*) subsequently, the animal is unable to get on its legs, displays transverse haemorrhagic streaks on the exposed rectal mucosa and the animal becomes off feed. With these symptoms, rate of mortality becomes high among calves.

Lesions

Though whole of the intestine is involved in coccidian infection in bovines, main lesions are concentrated in regions of small intestine. The mesentric blood vessels become distended and the intestinal coils appear extensively congested. On opening the intestine, there is catarrhal enteritis with reddening of the mucosa covered with excessive mucus. The intestinal contents are watery. There is a general swelling and punctiform haemorrhage spread over large areas. The coccidial lesions, as observed grossly, are typical and harbour a variety of developmental stages. These, mostly observed in the region of jejunum and ileum, are as follows :

(*i*) White pin-point to pin-head sized bodies lying superficially or deep in the mucosa. These are developing and mature *giant schizonts* (Fig. 10).

(*ii*) Irregularly margined but milky-white superficial areas of about 1-2 mm size. These are due to numerous immature and mature *large epithelial schizonts* (Fig. 11).

(*iii*) Slightly raised and small whitish patches of an oedematous character. These are due to collections of *gametocytes* and *oocysts* with *small epithelial schizonts* (Fig. 12).

(*iv*) Congested and oedematous areas with granular feature from aggregations of minutes white structures—the immature and mature microgametocytes of *Eimeria auburnensis*.

(*v*) Slightly raised, whitish, opalescent areas of 3-6

mm in diameter and with congested periphery revealing superficially the feature of white papillomatous lesions. This type of lesion harbours characteristic gametocytic and oocystic stage of *Eimeria bareillyi*.

The portions of large intestine also show lesions of severe catarrhal enteritis filled with fluid contents with shreds of blood and fibrinous clots. In paracute cases, rectum may also be involved in addition to caecum and colon. Such lesions harbour both asexual and sexual stages of coccidia viz. *Eimeria zuernii*.

Diagnosis

Diagnosis is based on the type of clinical symptoms especially diarrhoea. In paracute cases, the oocysts may rarely be seen in faeces particularly in the early phase of the disease. In sub-acute or chronic cases, large number of oocysts may pass out alongwith semisolid stools. On postmortem, the gross examination of the intestinal lining may reveal various lesions described above. Microscopic examination of fresh scrapings from these characteristic lesions and their stained preparations including sections would show different types of endogenous stages—schizonts, male and female gametogonic stages and oocysts in isolation or in groups.

Control

General hygienic conditions are essential for keeping the animals healthy as coccidiosis is mainly stall fed disease. Regular routine of investigation at large herds helps in early diagnosis of the disease for the immediate proper treatment of clinical and sub-clinical cases and their segregation from healthy stock. Though complete eradication of coccidia from domestic stock is neither practicable nor advisable, it is essential to put maximum efforts in providing clean water and feed with uninfected surroundings and grazing fields. Young calves should never be introduced directly into adult stock. The heavy intake of sporulated (infective) oocysts can be reduced by avoiding gross contamination of food and water and by washing floors and walls of byres at regular intervals. Bedding and surroundings may be sterilized by 1.25 percent sodium hypochloride or 0.5 percent cresol or phenol or by fumigation with formaldehyde. The enclosures, including byres and yards, should be kept dry as far as possible.

Besides coccidiosis, another protozoan disease 'balantidiosis' has been reported from buffaloes in India (Priestly, 1944; Biswal and Kanugo, 1959; Patnaik, 1965; Behera, 1967).

Treatment

Treatment of coccidiosis in cattle usually becomes difficult because the signs of the disease do not become noticeable until it is far advanced. Information on treatment of the disease in buffalo and cattle in India is meagre (Patnaik, 1965; Roychoudhury, Chakrabarty, and Dutta, 1970). Coccidiostats in vogue are of limited use as these have no effect on the sexual developmental stages which are responsible for many intestinal lesions. Of the many anticoccidial drugs tried with variable results, some have been found effective in controlling coccidiosis in cattle viz. Sulphamezathine, Sulphaquinoxaline, Sulphamerazine, Amprolium, Mepacrine hydrochloride, Pyrimethamine dapone and accestarsal, Lincomycin hydrocheloridc, Quixaline etc. Of all these, it appears that Amprolium is the most useful against two most commonly and pathogenic species, *E. zuernii* and *E. bovis*. The details about the doses etc. can be had from compilations on the subject (e.g. Pande, Bhatia, and Chauhan, 1972; Hammond, 1964; Pellerdy, 1974).

References

1. Behera, K.P. 1967. *Indian Vet. J.* **44** : 1080-81.
2. Bhatia, B.B. and Pande, B.P. 1967. *Acta Vet. Acad. Sci. Hung.* **17** : 351-157.
3. Bhatia, B.B., Pande, B.P., Chauhan, P.P.S. and Arora, G.S., 1968. *Acta. Vet. Acad. Sci. Hung.* **18** : 115-33.
4. Bhatia, B.B. Pande, B.P., Chauhan, P.P.S. and Garg, R.K. 1971. *Indian J. Anim. Sci.* **41** : 549-54.
5. Biswal, G. 1948. *Indian Vet. J.* **25** : 36-38.
6. Biswal, G. and Kanugo, K., 1959. Proc. 46th Indian Sci. Congr. Pt. III. p. 462.
7. Couper, H. 1927. *Indian Vet. J.* **4** : 56-64.
8. Cranti, V. 1956. *Profilassi* **29** : 133-41.
9. Guillon, J.C., Palisse, M. and Toueas, L. 1959. *Bull. Acad. Vet. Fr.* **32** : 513-25.
10. Hammond, D.M. 1964. *Coccidiosis* of *Cattle, Some Unsolved Problems.* 30th Faculty Association, Utah State University, Logan, Utah pp. 38.
11. Lotze, J.C. 1959. *J. Parasit.* **45** : Suppl. 40.
12. Malikyan, E.L. 1955. *Trud. armyan. nauchnoisolad. Vet. Inst.* **8** : 88-91.
13. Mandal, A.K. 1980. *Coccidia* and *Coccidiosis* of *Poultry and Farm Animals of India. Tech. Mono. 5. Zool.* Survey of India, Calcutta, p. 87.
14. Naik, R.N. 1927. *Indian Vet. J.* **4** : 137-39.
15. Pande, B.P., Bhatia, B.B. and Chauhan, P.P.S. 1971. *Indian J. Anim. Sci.* **41** : 151-54.
16. Pande, B.P., Bhatia, B.B. and Chauhan, P.P.S. 1972. *Coccidiosis in Domestic Ruminants-A*

Major Problem. Bulletin No. III. Para. Unit. U.P. Coll. Vet. Sci. & A.H., Mathura, p. 14.
17. Pande, B.P., Bhatia, B.B., Chauhan, P.P.S. and Garg, R.K. 1968. *Indian J. Vet. Sci.* **58** : 449-54.
18. Patnaik, M.M. 1965. *Indian Vet. J.* **42** : 186-91.
19. Patnaik, B. 1965. *Indian Vet. J.* **42** : 624-32.
20. Patnaik, M.M. and Pande, B.P. 1965. *Indian J. Vet. Sci.* **35** : 33-46.
21. Pellerdy, L.P. 1974. *Coccidia and Coccidiosis*, Verlag Paul Parey, Berlin and Hamburg. p. 959.
22. Priestley, F.W. 1944. J.R. Army, Vet. Cps. **16** : 129. (Cited by Mohan, 1968).
23. Rao, S.R. and Hiregaudar, L.S. 1953-54. *Bombay Vet. Coll. Mag.* **4** : 24-28.
24. Roychoudhury, G.K., Chakrabarty, A.K. and Dutta, B. 1970. *Indian Vet. J.* 47 : 265-67.
25. Shastri, U.V. and Krishnamurthy, R. 1975. *Indian J. Anim. Sci.* **45** : 46-47.
26. Shastri, U.V., Krishnamurthy, R. and Ghaffor, M.A. 1974. *Indian Vet. J.* **51** : 301-02.
27. Shastri, U.V., Krishnamurthy, R. and Ghaffor, M.A. 1979. *Indian J. Anim. Res.* **7** : 23-25.
28. Shumard, R.F. 1957. *J. Amer. Vet. Med. Assoc.* **171** : 559-61.

3. Fascioliosis

It is an important and serious disease of liver of domestic ruminants including water buffaloes in India. The National Commission on Agriculture reports that fascioliosis is widely prevalent in India and is one of the important parasitic diseases of bovine and bubaline population particularly in marshy and water-logged areas of the plain and hilly tracts.

Etiology

Two main species of large or giant-sized flukes, *Fascioa hepatica* and *F. gigantica* are the parasites of liver and bile ducts of these animals—the latter one being most commonly encountered.

Epidemiology

Though, the disease is widely distributed in the country but its prevalence is more in regions where climate is damp and having low lying and swampy areas with high rainfall and where natural water reservoirs are ample for the proliferation of snail vectors. The common species of snail involved in the life-cycle of these flukes in India is *Lymnea auricularia*. A high incidence of liver fluke disease is observed throughout Himalayan belt and foot of the hills, Tarai region and eastern zone than in other parts of the country. But for certain stray reports (Roy, 1954; Ahluwalia *et al.*, 1966), no systematic studies have been conducted in different parts of the country particularly on its epidemiological aspects.

Young calves are very susceptible to fasciolosis resulting in high rate of mortality. Animals become infected by ingestion of large number of the metacercariae encysted on grass blades and other plants in and around ponds, ditches or by swallowing the cercariae, settled at the bottom, alongwith water. After intake of large number of metacercarial stages, the young flukes reach the liver and bile ducts through intestinal mucosa in two ways. After penetration of the intestinal wall, the young flukes either enter the portal vein via intestinal veins or come directly into the peritoneum from where they reach the liver tissue and finally in the intrahepatic bile ducts. This migration may take several weeks.

The clinical picture of the disease depends mainly on the number of infective stages ingested by the animal at a time or repeated infections in a short time. Only in very heavy infection of young susceptible animals acute conditions appear, otherwise chronic form of fascioliosis is most commonly seen. The young flukes burrow in the liver and eat paranchyma as they pass through and cause extensive damage than the adult flukes do after they have reached the bile ducts. Traumatic hepatitis is produced by the migration of a large number of immature flukes which also suck large quantity of blood. There may be poisoning by their toxins and toxic metabolites. In acute condition, this results in rupture of the liver capsule with haemorrhage into the peritoneal cavity. This becomes the cause of sudden death of the animal. The dead animal shows discharge of frothy blood from the nostrils and also passing out blood from the anus. This needs be distinguished from the symptoms of anthrax. In chronic form, there is progressive biliary cirrhosis resulting ultimately in a hard fibrotic and enlarged liver in which the bile ducts are seen thick, fibrous and often calcified in older cases giving an appearance of rope-like or pipe-like structures called "Pipy liver". In massive infections, there may be blockade of the bile ducts leading to jaundice and other icteric disorders. There is hepatocellular failure and obstructive biliary cirrhosis resulting in cholestasis. The animals show many digestive disturbances, alternate diarrhoea with constipation, emaciation, loss of weight, hydremia, dullness, etc. Shalash and Salama (1962), observed that the fascioliasis was associated with an increase in size of ovaries, corpora lutea and follicle and an increase in the proportion of functioning ovaries of both buffaloes and cows slaughtered. A large number of animals suffer from this wasting disease thereby causing extensive economic loss.

Diagnosis

By finding eggs in the faeces, symptoms, lesions and recovery of flukes in postmortem cases.

Treatment

Among others, Roy and Reddy (1969, 1971), Roy and Shukla (1971), Ramanna and Rao (1973), Kumar *et al* (1980), Maru *et al* (1980) tried the following fasciolicidal drugs in naturally infected cases of fascioliosis/fascioliasis in buffaloes with partial to 100% effectiveness in elimination of liver flukes.

1. Hexachlorethane	eg.	"AVLOTHANE" ICI
2. Hexachlorothene	eg.	"DISTODIN" Pfizer
3. Oxyclozanide	eg.	"ZANIL" ICI
4. Nitroxynil	eg.	"TRODAX" M & B
5. Disophenol	eg.	"ANCYLOL" Cyanamid.
6. Rafoxanide	eg.	"FLUKANIDE" MSD

There are other anthelmintics marketed under different trade names and are used against *Fasciola* in all other species of animals eg. 'Hetol' (Hoechst), Niclofolan, Difluoro-tetrachlorethane, "Clioxanide (Parke Davis), Carbon tetrachloride or its substitude Fasciolin etc.

References

1. Ahluwalia, S.S., Pande, B.P. and Bhatia, B.B. 1966. *Some of the Chief Pathogenic Helminths of Adult Buffalo in Uttar Pradesh.* Bull No.

II. Para. Unit. Coll. Vet. Sci. & A.H. Mathura p. 3.

2. Animal Health Year Book, 1978. *Animal Health Service, Animal Production and Health Division, F.A.O. W.H.O.* eg. Konigshofer, H.O.).
3. Kumar, M., Maru, A. and Pachauri, S.P. 1980. *Indian Vet. Med. J.* (In Press).
4. Maru, A., Kumar, M. and Pachauri, S.P. 1980. *Indian J. Anim. Sci.* (In Press).
5. National Commission on Agriculture (Report). 1976. Govt. of India, Ministry of Agric. & Irrigation. Vol. **7** : 466.
6. Ramana, B.C. and Rao, P.R. 1973. *Indian Vet. J.* **50** : 936-39.
7. Roy, S.C. 1954. Proc. 41st *Indian Sci. Congr. Pt.* III. (Abstr.) 224-25.
8. Roy, R.M. and Reddy, N.R. 1969. *Vet. Red.* **85** : 85-87.
9. Roy, R.M. and Reddy, N.R. 1971. *Cashiers de medicine Veterinarie* **40** : 337-43.
10. Roy, R.M. and Shukla, S.S. 1971. *Trop. Anim. Hlth. Product.* **3** : 26-31.
11. Shalash, M.R. and Salama, A. 1962. *Proc. IV. Inter. Congr. Anim. Reprod.* The Hague, 1961. 3 : 693-95 (*Vet. Bull.* **33** : 1972).

2

BLOOD VESSELS, CONNECTIVE TISSUES, LIGAMENTS, BODY CAVITY ETC.

1. Nasal Schitosomiosis

Etiology

'Nasal granuloma' (snoring disease) is an important and widely prevalent parasitic disease of cattle in the country caused by *Schistosoma nasale*. Buffalo, sheep and goat also harbour this infection.

Incidence

There are numerous reports on this disease from different states viz. Assam, Bengal, Bihar, Orissa, Madhya Pradesh, Maharashtra, Tamil Nadu, Karnataka, Andhra Pradesh, Punjab, Kerala and Uttar Pradesh. Many workers have realised that this disease has direct bearing on the production as Naik (1942) stated that working animals were rendered unfit for work when affected with nasal granuloma and Varma (1954) reported upto 60% infection in cattle in some districts of Bihar. According to Rao and Devi (1971, 1971a), though more of buffaloes carried *S. nasale* infection, cattle were found to excrete 2-3 times more eggs than buffaloes. Likewise, Rajmohan and Peter (1975), Muraleedharam *et al* (1976) and Philegaonkar *et al* (1977) studied the incidence of *S. nasale* among cattle and buffalo and found no gross visible lesions in the nasal mucosa in buffaloes. Systematic surveys are still needed to be undertaken in other States of the country in different geoclimatic condition among buffalo population so that the husbandry practices may be reoriented if buffalo only acts as a reservoir for *S. nasale* infection for other susceptible hosts.

Lesions

The blood flukes are seated in veins of nasal mucosa causing continuous irritation, mucopurulant discharge, proliferation of epithelial lining and fibrous tissue, formation of numerous abscesses containing eggs of the worm. In chronic cases, cauliflower like growth appears inside nasal cavity blocking the free air passage. There is continuous discharge from the nostrils, eyes and there is dyspnoae and snoring. The disease adversely affects the health and production of infected animals.

Diagnosis

By lesions and finding eggs in nasal discharges.

2. Onchocerciosis

Etiology

Onchocerciosis with three distinctly recognised species *Onchocerca armillata* parasitizing wall of het bulbous and thoracic aorta. *O. gutturosa* parasitizing ligamentum nuchae and *O. gibsoni* in nodules and worm nests in the subcutaneous connective tissues in buffaloes and cattle is widely prevalent in our country. *O. armillata* has additionally been reported from sheep, goat and equines (Bhatia and Sood, 1959; Bhatia, 1960; Katiyar and Pande, 1964).

Lesions

In *O. armillata* infection of the aorta, the internal surface at the affected sites appears deeply corroded, exhibits complex sinuous tunnels of brownish colour running beneath the intimal layer and slightly elevated nodules of varying sizes of bluish colour and lying interspersed in the area. Nodule, on teasing, may yield a coiled male worm and the anterior part of the female worm with its main body extending back into the sinuous tunnels (Fig. 13). From its burrowing activities, the parasite is responsible for destruction of muscle fibres and elastic laminoe of the aortic wall leading to fibrosis and even calcification at places. As *O. armillata* is also found in other host species, buffalo may be an important reservoir in the spread of this species through appropriate transmitting agents in the areas which need be known for effective control measures (Chauhan and Pande, 1972a). Among other reports, Singh and Singh (1970) studied gross and microscopic pathology of *O. armillata* infection in buffalo and concluded that the infection must significantly affect the blood circulation. Shastri (1973) reported its incidence in buffalo in Maharashtra State. Singh *et al* (1973) found, beside *O. armillata*, aortic wall infected with *Eleophora poeli* in 51 of 256 buffaloes at Bareilly with umblicus-like corrugated lesions, migratory tract and nodulation. Prasad and Bhalla (1977) reported the prevalence of parasitic aortitis due to *F. poeli* in 19% water buffalo.

O. guttuorsa affects the ligamentum nuchae of which the intermuscular tissues show intricately coiled worms deeply amongst muscular bundles. On palpation, nodules of different sizes, lying deep and/or superficially, can be felt. On incision, circumscribed whitish and calcified areas harbouring degenerated worms are observed. Such lesions occur on the head, withers, neck and shoulders. Marked cellular infiltration with developing fibrosis could, at places, be observed. There may also be some degenerative changes around arrested worms. Microfilariae are seen in and around the lesions. The older lesions show hard nodules and, on incision, reveal chalky appearance (Pande and Ahluwalia, 1964; Chauhan and Pande, 1972a). Shastri (1973) reported its incidence in buffaloes in Maharashtra State.

O. Gibsoni nodules, both fibrous and calcarious, occur in various subcutaneous tissues of legs and other parts of body. The fibrous nodules usually yield both adult worms and microfilariae in the peripheral regions and the overlying skin (Ahluwalia *et al.* 1966).

Diagnosis

By finding microfilariae in blood smears and recovery of worms from lesions.

3. Setariosis

Etiology

Generally, species of *Setaria* are found in peritoneal cavity of variety of hosts. Five different species of *Setaria* are known from Indian buffalo of which *S. labiatopapillosa* is found most commonly. Its pathogenic significance is only realised when the worms are located in such ectopic foci as upper eye-lid, liver, heart, pericardium and pleura, sub-cutaneous cysts or fascia, oviduct, kidneys, urinary bladder etc. Additionally, a greater significance is attached in view of the role of *S. digitata* in cerebrospinal nematodiasis in sheep, goat and horse and cerebrosetariasis in bullocks. A few cases of prenatal infection are also on record. Pachauri (1972) reported cerebrospinal nematodiasis in a murrah buffalo due to *S. cervi*.

A number of cases of ectopic foci of setariid worms in buffaloes including the adult female worms embedded under the serosa in the region of small intestine and immature worms in the sub-serosal coat. Perhaps, the association of adult female worms with serosal cost of intestine might be for the deposition of microfilariae which could easily be trrnsported to general circulation (Chauhan and Pande, 1980).

Diagnosis

Same as in Onchocercosis.

Microfilariasis is identified as a disease entity with certain manifestations occurring due to the presence of microfilariae in blood circulation of the host. Mukherjee (1965) reported a high incidence of microfilarial incidence in the blood of buffaloes. Singh *et al* (1973) made some clinical observations in buffaloes carrying *Seteria cervi* and *Onchocerca armillata* infections displaying circling movements, reduced milk yield, swelling of the legs and raised boay temperature. They also made haematological studies. Kumar (1980) observed overall incidence of microfilariasis in buffaloes above 23% more in lactating buffaloes than non-lactating ones and least in hiefers. The usual symptoms in microfilariasis in buffaloes reported by different workers are : anorexia, weakness, rise in temperature, congestion of conjunctiva, discharge of watery exudate from nostrils, mucopurulent discharge from eyes, reduction in milk yield or even complete cessation of milk in buffaloes in late lactating stage, unable to stand on hind legs, and may also show eczematous dermatitis and weaping wounds over the body.

4. Parafilariosis

Etiology

Alwar and Lalitha (1958) reported parafilariasis in buffalo for the first time in India. Subsequently, Srivastava and Dutt (1959) observed a large number of cases of parafilariasis in buffaloes on a farm in Bihar and identified incriminated filarial nematode through examination of a single female as a new species *Parafilaria sahaii*. Later, Patnaik and Pande (1963), who also examined worms from lesions of buffaloes, considered that the species in buffaloes was also *Parafilaria bovicola* parasitizing normally cattle.

Lesions

The nodules, of about 2-3 cm in diameter, develop on the neck and body mostly in adult animals. This is a seasonal disease which is most prevalent during wet monsoon summer season. Only during sunny days, blood oozes out of the nodules which carry eggs containing microfilariae released from the protruding female worms. Both eggs and hatched out microfilariae are easily detectable in freshly oozing blood. Data on the incidence and pathogenesis of parafilariasis in Indian buffaloes are still lacking.

Diagnosis

As in onchocersis.

5. Spirocercosis

Etiology

Besides *O. armillata* infection of the aortic wall, infection with juveniles of *Spirocerca lupi* has also been recorded in addition to its occurrence in aortae of goat and equines (Pande *et al*, 1961; Katiyar and Pande, 1965). The tissue reactions in these abnormal hosts against the juveniles of the common oesophageal tumour worm of dog are extensive.

Lesions

In aorta of buffalo, the juveniles remain embedded inside the wall and, in some cases, partly hanging in the lumen. Lesions are seen irregularly margined and different sized outgrowths with ulcerations, haemorrhagic spots and blood clots. If and when such atheromatic lesions become markedly prominent, partial or complete occlusion of the blood supply, formation of emboli, from blood clots and even rupture of the attacked blood vessels could result (Chauhan and Pande, 1980). Infection of such abnormal hosts occurs by accidental ingestion of dung beetles harbouring the infective stages of *S. lupi*.

Diagnosis

By finding worms in the lesions during postmortem examination.

6. **Thelaziosis** (eye worm infection)

Etiology

'Thelioziosis', an eye worm infection, has recently been realised as an important disease condition responsible for considerable economic loss among buffaloes and cattle population in India. The importance of this disease has been emphasized by workers abroad who have reported its incidence as high as 92% in endemic areas. *Thelazia rhodesii* was the only known species involved in thelaziasis in buffaloes and cattle in India till the work of Pande *et al* (1970) who found involvement of three species, *T. rhodesii*, *T. skrjabini* and *T. gulosa*—the last two were reported for the first time. The authors gave an adequate description and histopathology and found *T. skrjabini* of most common occurrence. Subsequent reports were of Venkatratnam *et al* (1969), Indra Devi *et al* (1970) and Bhopale *et al* (1970, 1971). According to Arora *et al* (1975), the highest incidence of eyeworm remained during the winter season (59.45%) than in rainy season (44.23%) and summer season (41.34%); the most frequent species was *T. skrjabini* (96.97%) followed by *T. rhodesii* (7.59%) and *T. gulosa* (3.36%) among positive cases. The incidence and intensity of *T. skrjabini* and *T. gulosa* was greater in winter followed by rainy and summer seasons. Mature worms of these species occurred in greater number in winter and immature ones in rainy season. The infection rate of *T. rhodesii* was highest during summer but its intensity was greater in winter when more mature worms occurred than in rainy season when immature worms were found in greater number.

Symptoms

The infection is generally associated with varying

grades of eye diseases due to localization of these nematodes in the inner canthus below the third eye lid, in the lacrimal duct or tear ducts or even in deeper sites. In living condition, the affected animals show symptoms of conjuctivitis, excessive lachrymation, mucopurulent discharge, abscess in eye lid, photophobia and, in severe cases, corneal opacity and keratitis leading to complete blindness. These manifestations ultimately result in emaciation and reduction in milk production due to constant annoyance and irritation caused by the worms. The animals show anorexia, anxiety and toss their heads about especially at the first stage of the disease. Secondary microbial invasion may complicate the condition. The worms also have a toxic effect.

Lesions

The eye worms reveal varying grades of pathological changes on histologic examination. Besides, petechiae at different points of the conjunctiva, collection of red blood cells, some degree of hyperthryphy of the lymphoid follicle, infiltration with a few eosinophils and desquamation of the epithelial lining (Figure 14). The worms invade the associated muscles through the surrounding areolar tissues. Further investigations done by Chauhan *et al* (1976) resulted in recovery of two adult male worms of *T. skrjabini* from vitreous humour in one of the eyes of a buffalo examined an unusual site of its occurrence. The same eye also yielded 11 male and 18 female specimens of the same species from the lacrimal duct. Occurrence of immature parafilarial worms in the anterior chamber of eye of a buffalo is also recorded (Chauhan, *et al*, 1974).

Diagnosis

By finding worms in eye wash or through mechanical

removal.

Control

As, theloziasis is a seasonal disease and is mainly prevalent during fly season, its incidence can be reduced substantially by elimination of the fly vectors (*Musca* sp. or related species) by use of fly-repellents and by keeping the animals and their surroundings clean and dry besides proper treatment of affected cases.

Other helminthic infections reported from Indian buffaloes are 'Syngamiasis' due to *Syngamus laryngeus* (Mudaliar and Alwar, 1947; Shah *et al*, 1960; Pathak, 1963 Bhattacharjee, 1966, Endrejat, 1964), and lungworm infection due to *Dictyocaulus viviparous* (Endrjat, 1964, Jain and Bandopadhyay, 1967) which need further investigations in different geo-climatic regions of the country for proper understanding of their prevalence and pathogenicity, if any. However, a recent report on the pathology of larynx due to *S. laryngeus* is available (Bhopale *et al*, 1982). They reported 3.6% of infection among buffalo calves. The larynx revealed congested mucous membrane, catarrhal exudate over the mucous membrane and attachment of the worms to the mucosa.

7. Stephanofilariosis

Etiology

Two distinct species of the genus, *Stephanofilaria*, *S. assamensis* and *S. zaheeri* are commonly found in many parts of the country causing 'hump sore' in cattle by the former and 'ear sore' in buffaloes by the latter species. Buffalo seems to be the only host for *S. zaheeri* causing characteristic lesions of ear sore in a prominent form of dermatitis. The worms are nearly of the same size of

S. assamensis but for slightly larger size of female worms. Ear sore lesions in buffaloes was first reported by Gopalakrishnan (1948) from Assam. The causative species was identified by Singh (1958). Subsequently, the disease was found to occur in various parts of the country particularly in eastern India eg. Bengal, Assam, Orissa, Bihar and also from Madhya Pradesh (Bhopale and Joshi, 1970; Agrawal, 1973).

Epidemiology

Mostly, adult buffaloes suffer from ear sore disease. Usually, the rate of infection increases with the advancement of the age of the animal. The infection rate increases upto the age of 4 years and thereafter it decreases. Maximum infection has been observed in animals of 2-4 years and 10-16 years of age. Lesions are usually not seen in calves below 5 months of age (Agrawal and Dutt, 1977a, Agrawal and Dutt, 1978). Patnaik and Khan (1972) briefly discusssd its incidence, distribution, pathology and economics in the country. Das *et al* (1975) reported incidence in Orissa and found active lesions mostly in the rainy season and quiescent ones in the months of January and February.

Lesions

The worms lie in and around hair follicles and sebaceous glands. The lesions are categorised as moderate, chronic and atypical (Agrawal, 1977). Moderate lesions show thickening, cracked skin with or without exudation while chronic lesions exhibit granulated, dry and hard skin covering whole of the inner surface of pinnah. Atypical lesions have scattered black spots. Pathology of ear sore in buffaloes has been studied by Patnaik (1964), Bhattacharjee and Das (1967), Patnaik

(1970) and Agrawal *et al* (1978). Generally, the gross lesions are characterised in three stages. In the beginning there are erythematous spots which coalesce at places sometimes with roughness of the skin. Subsequently, the lesions show thickened, reddish cracked skin with serous exudation leading to crust formation. Oozing of fresh blood may be seen at some spots. Such lesions may show depigmentation without hairs. In later stages of infection, there is encrustation. granulation, dryness, without exudation, covering nearly whole of the inner surface of pinnah. Parakeratosis, acanthosis and cut worms are generally seen on histologic examination in epidermis and dermis. The inflammatory changes occurring in the lesions are primarily because of filarial worms and not due to secondary bacterial invasion.

Diagnosis

By finding microfilariae and/or worms in the lesions.

Treatment

Nasal Schistosomiasis

There are a number of antimonials i.e. tartar emetic, anthiomaltine, antimosan and neguvon used in the treatment of bovine nasal schistosomiasis. Their chemotherapeutic values in buffaloes are not yet known. Likewise, other antimonials eg. Sodium antimony dimercapton succinate and Stibophen (Foudin) and nonantimonials eg. Ambihar (Niradizole) and Miracil-D (Lucanthone), commonly used in human schistosomiosis, need also be tried for assessing better therapeutic results in both bovine and bufaline nasal and visceral schistosomiosis. Bhatia and Rai (1976) tried Lithium antimony thiomalate (Anthiomaline, M & B), Stibophen (Antimosan, Bayer),

Potassium antimony tartarate (Tartar emetic) and Niridazole (Ambilhar 'Ciba') against clinical cases of nasal granuloma in cattle and found tartar emetic and Antimosan being most effective in about a month's time.

Onchocerciosis

No treatment is known for the disease.

Setariasis & Microfilariosis

Patnaik and Roy (1968) reported that 6% Neguvon in linseed oil was effective in 7-10 days in *Setaria* infection in buffaloes and goats. Gupta and Pachauri (1981) tried Anthiomaline and Caricide in combination for a period of 10 days alongwith supportive therapy with Calcium Magnisium Borogluconate and liver extract and Vitamins (Belamyl). The animals appeared to show remarkable clinical improvement and no microfilariae were found on blood examination for 3 successive days.

Parafilariosis

Sahai *et al* (1965) treated the bovine cases with 2% of Sodium antimony tartarate given sub-cutaneously twice a week for 3 weeks. Other drugs, commonly used in bovine parafilariosis, are Neguvon, Antypol, Anthiomaline, and Antimosan.

Spirocercosis

No treatment is known.

Thelaziosis

Chauhan (1977) suggested that all the positive cases

of eyeworm infection, with or without clinical manifestations, may be treated simultaneously in the entire herd. The eyes may be rinsed with a strong jet of 2 or 3 percent aquous solution of boric acid by a rubber bulb or a large syringe. Most of the parasites are removed by first irrigation. Eye-irrigation may be repeated twice or more with intervals of 5 to 6 days. Wiping of the conjunctival sac with small cotton wool tampons wetted in a solution of boric acid is useful in removing the parasites mechanically. The irrigation of eyes has also been recommended with 50-75 ml of aquous solution of 0.5 percent iodine and 0.75 percent pottasium iodide or with solutions of 2-3 percent of boric acid, 0.05 percent of mercuric chloride, 1 in 2,000 iodine, 0.5 percent lysol or 1 in 500 diethyl-carbamazine. Antibiotic opthalmic ointments and solutions should also be used in conjunction with above treatment to combat secondary microbial infections. Application of Butacaine sulphate 20 per cent and Nitromisol 1:3,000 opthalmic ointment to the conjunctival surface relieves the pain.

Proper treatment of the affected animals, particularly during the late winter season (January to March), when the animals carry adult stage of the worms, minimises the chances of propagation of the infection in the following months of fly season.

Stephanofilariosis

Gopalakrishnan (1951) and Ahmed (1961) used tartar emetic ointment (1:25); Raghavachari and Reddy (1957) observed temporary response with tartar emetic and anthiomaline; Mishra (1969) claimed success by two applications of 5% formalin; or one application of 10% formalin; Das *et al* (1975) treated 34 out of 48 affected cases of ear sore in buffaloes by application of Supona 20 or Sumithion used as 4%

ointment for 30 days; and Agrawal and Dutt (1977) observed that Malathion (6% ointment) gave the best result with a cure of 58.3% and improvement of 41.7% followed by acrifilavin (2% ointment) with cure rate of 41.7% and improvement rate of 25%.

References

1. Agrawal, M.C. 1973. M.V. Sc. Thesis, JNKVV, Jabalpur.
2. Agrawal, M.C. 1977. *Orissa Vet. J.* **11** : 122.
3. Agrawal, M.C. and Dutt, S.C. 1977. *Indian Vet. J.* **54** : 77-79.
4. Agrawal, M.C. and Dutt, S.C. 1977a. *Indian Vet J.* **54** : 497-498.
5. Agrawal, M.C. and Dutt, S.C. 1978. *Indian J. Anim.* **48** : 232-234.
6. Agrawal, M.C.; Vegad, J.L. and Dutt, S.C. 1978. *Indian J. Anim. Sci.* **48** : 261-265.
7. Ahluwalia, S,S.; Pande, B.P. and Bhatia, B.B. 1966. *Some of the Chief Pathogenic Helminths of Adult Buffalo in Uttar Pradesh. Bull.* No. II. Para. Unit. Coll Vet. Sci. & A.H. Mathura p. 3.
8. Ahmed, Z. 1961. *Indian Vet. J.* **38** : 257.
9. Alwar, V.S. and Lalitha, C.M. 1958. *Madras Vet. Coll. Ann.* **16** : 15-19.
10. Arora, G.S.; Chauhan, P.P.S.; Agarwal, R.D. and Ahluwalia, S.S. 1975. *Indian J. Anim. Sci.* **45** : 953-57.
11. Bhatia, B.B. 1960. *Indian Vet. J.* **37** : 394-97.
12. Bhatia, B.B. ahd Sood, S.M. 1959. *Curr. Sci.* **28** : 490-91.
13. Bhatia, B.B. and Rai, D.N. 1976. *Indian J. Anim. Res.* **10** : 43-44.
14. Bhattacharjee, M.L. 1966. *Indian J. Anim. Hlth.* **5** : 143-45.

15. Bhattacharjee, M.L. and Das, D.D. 1967. *Indian Vet.J.* **44** : 400-402.
16. Bhilegaonkar, N.G.; Babras, M.A. and Sardey, M.R. 1977. *Abstr. National Congr. Parasit. Baroda.* 24-26.
17. Bhopale, K.K. and Joshi, S.C. 1970. *Orissa Vet. J.* **5** : 19-21.
18. Bhopale, K.K.; Joshi, S.C.; Jain, P.C. and Kamalapur, S.K. (1970). *Indian Vet. J.* **47** : 564-66.
19. Bhopale, K.K.; Joshi, S.C. and Kamalapur, S.K. 1971. *Iadian Vet. J.* **48** : 346-48.
20. Bhopale, K.K.; Joshi, S.C. and Shah, H.L., 1982. *Indian Vet. J.* **59** : 396.
21. Chauhan, P.P.S. 1977. *Indian Farming* **27** : 29.
22. Chauhan, P.P.S. and Pande, B.P. 1972. *Indian J. Anim. Sci.* **42** : 809-13.
23. Chauhan, P.P.S. and Pande, B.P. 1980. (Unpublished).
24. Chauhan, P.P.S., Arora, G.S. and Ahluwalia, S.S. 1974. *J. Helminth.* **48** : 289-93:
25. Chauhan, P.P.S., Arora, G.S., Agrawal, R.D. and Ahluwalia, S.S. 1976. *Indian J. Anim. Sci.* **46** : 152-53.
26. Das, C.M.; Tripathy, S.B. and Misra, S.K. 1975. *Indian J. Anim. Sci.* **45** : 949-952.
27. Endrejat, E. 1964. *Indian Vet. J.* **41** : 538-42.
28. Gopalakrishnan, N.R. 1951. *Indian Vet. J.* **38**:257.
29. Gopalakrishnan, V.R. 1948. *Indian J. Vet. Sci.* **18** : 227.
30. Gupta, R.K. and Pachauri, S.P. 1982. *Agric. Sci. Digest* **2** : 65-67.
31. Indra Devi, T.; Padmavathi, P.; Venkatratnam, A. and Ramkrishna Reddy, P. 1970. *Indian Vet. J.* **4** : 275-77.

32. Jain, S.K. and Bandyopadhyay, A.C. 1967. *Indian Vet. J.* **44** : 661-64.
33. Kumar, M. 1980. M.V. Sc. Thesis, Department of Medicine, Coll. Vet. Sci., G.B. Pant Univ. Agric. & Tech., Pantnagar.
34. Katiyar' J.C. and Pande, B.P. 1964. *Agra. Univ. J. Res.* (*Sci*). **14** : 227-34.
35. Katiyar, J.C. and Pande, B.P. 1965. *Agra. Univ. J. Res.* (*Sci*). **14** : 151-56.
36. Mishra, S.P. 1969. *Indian Vet. J.* **46** : 632.
37. Mukherjee, R.P. 1965. *Indian Vet. J.* **42** : 131.
38. Muraleedharan, K.; Kumar, S.P.; Hegde, K.S. and Alwar, V.S. 1976. *Mysore J. Agric. Sci.* **10** : 105-17.
39. Pachauri, S.P. 1972. *Indian J. Anim. Res.* **6**:17-19.
40. Pande, B.P. and Ahluwalia, S.S. 1964. *British Vet. J.* **120** : 132-37.
41. Pande, B.P.; Rai, P. and Bhatia, B.B. 1961. *J. Parasit.* **41** : 951-52.
42. Pande, B.P.; Chauhan, P.P.S.; Bhatia, B.B. and Arora, G.S. 1970. *Indian J. Anim. Sci.* **40** : 330-45.
43. Patnaik, B. 1970. *Workshop on Stephanofilaria dermatitis.* Indian Veterinary Research, Institute, Izatnagar.
44. Patnaik, B. and Khan, M.H. 1972. *Indian Farming* **22** : 127-28.
45. Patnaik, B. and Roy, S.P. 1968. *Indian Vet. J.* **38** : 55.
46. Patnaik, M.M. 1963. *Indian Vet. J.* **40** : 272-74.
47. Patnaik, M.M. 1964. *Indian Vet. J.* **41** : 462-66.
48. Patnaik, M.M. and Pande, B.P. 1963. *Indian Vet. J.* **40** : 128-33.
49. Prasad, M.C. and Bhalla, N.P. 1977. *Indian Vet. J.* **54** : 97-101.

50. Ragavachari, K. and Reddy, A.M.K. 1957. *Indian Vet. J.* **34** : 321.
51. Rajmohanan, K. and Peter, C.T. 1975. *Kerala J. Vet. Sci.* **6** : 94-100.
52. Rao, P.V.R. and Devi T.I. 1971. *Purdue. Univ. Studies* 303-307.
53. Rao, P.V.R. and Devi, T.I. 1971a. *Indian J. Anim. Hlth.* **10** : 185-88.
54. Shai, B.N.: Singh, S.P. and Srivastava, V.K. 1965. *Indian Vet. J.* **42** : 881-83.
55. Shastri, V.V. 1973. *Indian J. Anim. Hlth.* **12** : 193-95.
56. Singh, D.V., Joshi, H.C. and Shivanani, G.A. 1973. *Philippine, J. Vet. Med.* **11** : 101-09.
57. Singh, K.P. and Singh, C.M. 1970. *Indian Vet. J.* **47** : 1135-37.
58. Singh, S.N. 1958. *J. Helminth.* **32** : 239.
59. Srivastava, H.D. and Dutt, S.C. 1959. *Curr. Sci.* **28** : 128-29.
60. Varma, A.K. 1954. *Indian J. Vet. Sci.* **24** : 11-33.
61. Venkataratnam, A., Padmavathi, P. and Indra Devi, T. 1969. *Indian Vet. J.* **46** : 261-62.

3

HAEMOPROTOZOAN INFECTIONS

1. Babesiosis

Buffalo has been reported to be a refractory host to *Babesia* infection. Lingard and Jenings (1904) (cited by Mohan, 1968) first reported larger piroplasm in buffaloes in India. Parthasarthi (1931) identified *Babesia bigemina* in the peripheral blood of 6 and in heart blood of 16 post-mortem cases of buffaloes in South India. Idnani (1938) and Narusu (1940) reported isolated cases of clinical babesiosis in buffaloes and incriminated *B. bovis* and *B. bigemina* in heart blood and peripheral blood films of buffalo-calves at autopsy examination. Mohan (1968), in his review, states that babesiosis is infrequent in buffaloes. The reason might be due to low prevalence of ticks on buffaloes and innate resistance, Roychoudhury and Gautam (1979) could not produce clinical disease experimentally in intact buffalo-calves though parasitaemia with fever was observed in spleenectomised buffalo-calves. Their experiment studies support the general belief that, under natural conditions, buffaloes are refractory to *Babesia* infection. Dwivedi *et al* (1979) reported two clinical cases of babesiosis in buffaloes due to *B. bovis* infection with symptoms of high temperature (105°–106°F), anaemia, anorexia dullness, depression, dsypncea, disinclination to move and severe haemoglobinurea. They also treated the cases with Berenil.

Symptoms

Buffaloes mostly suffer from latent symptomless but fatal infection in new borns may exist. The resistance of calves disappears in 9-12 months and babesiosis may cause higher mortality in adult animals than in young calves. Incubation period is 1-2 weeks for *B. bigemina* and 4-10 days for *B. bovis*.

There is sudden rise of temperature to 41-45°C which lasts from a few days to a week or more. In acute form of the disease, the blood shows parasitaemia upto 75 percent resulting in hypochromic normocytic anaemia and haemoglobinurea. There may be gradual appearance of reticulocytes and rise in mean corpuscular volume and the anaemia becomes macrocytic. There is damage of internal organs by anaemic anoxia particularly of kidney and liver. Appetite is either subdued or there may be complete loss of appetite. There is staggering gait, suppressed rumination, discharge from the eyes, trickling of saliva. Respiration and heart rates are increased. The mucous membrane and conjunctivae become brick-red but soon change to extreme pallor of severe anaemia. In the terminal stages, there is severe jaundice and the urine becomes dark red to brown in colour and produces a froth. Death may occur within

4-8 days of infection in acute cases. In chronic cases, there may not be haemglobinurea. The animal becomes dull, listless and passes yellowish brown faeces. The remay be diarrhoea or constipation and weakness of hind quarters.

Lesions, on post-mortem, consist of sub-cutaneous and intramuscular oedema with icterus. The fat is gelatinous and yellow. The blood is thin and watery. The spleen is markedly enlarged, soft and dark. Liver is enlarged, pale and yellowish with distended gall bladder. In cerebral form, there is perivascular, perineuronal and interstitial oedema throughout the brain and spinal cord.

2. Theileriosis

There is scanty information on the frequency and pathogenicity of theileriosis among buffaloes in India. Earlier, according to Mohan (1968), Lingard and Jennings in 1904 recorded a small piroblasm in the blood of buffaloes and cattle in India having ovoid and bacillary forms. This was later confirmed to be *Theileria mutans*. In general, *Theileria* parasites aie less frequent in buffaloes than in cattle like *Babesia* (Mohan, 1968). As mentioned by Sahai *et al* (1962), Ramiah recorded *T. annulata* infection in Indian buffaloes.

Marutyan (1955) studied experimental *T. annulata* infection in buffaloes. The disease took either an acute or a latent form. Symptoms and lesions were less pronounced than in cattle. The infection could be transmitted by means of *Hyalomma detritum* and *H. anatolieum.*

Chhabra *et al* (1972) recorded a natural case of theileriasis in a buffalo in Punjab. They observed high body temperature (104°—107°F), dullness, emaciation, discharge from eyes and mild constipation. Blood smears revealed the presence of oval and rounded ringshaped bodies in the erythrocytes. Koch's blue bodies, however, could not be seen in smears.

Burridge and Odeke (1973) experimentally established *T. lawrenci* infection in two buffaloes by inoculation of stabilate material. The animals died after 15 and 16 days of exposure. They could detect only macro-schizonts in the peripheral lymph nodes 8 days later. Dhar *et al* (1973) also produced the disease experimentally in spleenctomised buffaloes with low parasitaemia without manifestation of clinical conditions.

Theileriosis, in general, is one of the most important tick-borne blood protozoan disease of ruminants. Two species viz. *Theileria annulata* and *T. mutans* are found to parasitize cattle and buffaloes in India. The former species could be fatal and the latter is responsible for benign cases. The parasites are characterised by a reproductive cycle resembling schizogony in lymphocytes and/or their stem cells and further developmental stages occur in erythrocytes and small, pleomorphic, nonpigmented bodies known as piroplasms and these are known to divide by binary fission. The piroplasm forms in the red blood cells of *T. annulata* are nearly indistinguishable from *T. parva* and most common forms occur as round, oval or ringshaped of about 0.5-1.5u in size. Besides, rod-shaped, commas and anaplasma-like organisms may also be found. As young as 2-3 days old calves have been reported to show clinical theileriosis (Sharma, 1970) and, therefore, intra-uterine infections may be a possibility. Mostly animals suffer from this disease during summer and rainy season which coincides with the high incidence of ticks from May to October. Raghavachari (1944) recognised theileriosis as a menace to calf rearing in India.

Transmission of theilerias has been established in India by at least four species of *Hyalomma* ticks viz, *H. anatolicum anatolicum* (Bhattacharyulu *et al*, 1972),

H. detritum (Gill, *et al*, 1974), *H. dromedarii* and *H. marginatum isacci* (Bhattacharyulu *et al*, 1975). The mode of transmission by these ticks is stage to stage or trans-stadial. The stage (larva or nymph), feeding on the infected animal, transmits the parasite to other animal through its next stage (nymph or adult). This proves that the disease can also be transmitted mechanically to susceptible animals by inoculating infective blood and tissue suspensions made from spleen, lymph nodes and liver of infected animals. Buffaloes are supposed to act as reservoirs but may also suffer.

Symptoms

Tropical theileriosis, with incubation period of 10-25 days, runs acute, subacute or chronic course. Clinical symptoms and lesions in buffaloes are less pronounced than in cattle. The disease may last for 3-4 days in the acute form or it may continue for about 3 weeks. There is marked rise of temperature reaching 40-41°C followed by depression, lachrymation, tumefaction of of eyelids and ears, nasal discharge, cessation of ruminations, diarrhoea with blood or mucus and swelling of the superficial lymph nodes. In fatal cases, there is rapid loss of body weight. In acute form, there is dyspnoea and death due to oedema of lungs. It is not always that the severity of the disease corresponds to the extent of parasitaemia. In severe infection, haemoglobinaemia, billirubinaemia and anaemia are features of the disease. Haemoglobinuria may occur. Jaundice is rare. Parasitaemia lasts for 5-54 days.

In the chronic form, there is intermittent fever, loss of appetite, marked emaciation, anaemia and icterus may persist for a variable long time.

Lesions

There is marked enlargement of lymphnodes, spleen and liver. Lungs are usually oedematous and kidney may reveal infarction. Mucous membranes become pale and petechiae. Abomasum and small intestine are inflamed and the former show punched necrotic characteristic ulcers of 2-12 mm in diameter with a zone of inflammation which are pathognomic.

Diagnosis is based on symptoms, microscopic examination of blood smears, biopsies from swollen superficial lymph nodes, serological tests viz. indirect haemagglutination test, indirect fluorescent antibody test, complement fixation test, immunodiffusion test, capillary agglutination test etc., and lesions on postmortem examination (Dhar and Gautam, 1977a, b, c, d).

3. Buffalo Malaria

Plasmodiam bubalis was first detected in the blood of buffalo by Sheather (1919). Hasan (1953) recorded an overall incidence in over one thousand buffaloes examined by him who noted a 72 hr periodicity of pyrexia in some buffaloes. According to Anon (1977), *P. bubalis* is seen occasionally in the blood of buffaloe especially when they are subjected to repeated bleeding for serum production or when their resistance is lowerd in other ways. The appearance of the parasite in peripheral blood, however, is not known to be accompanied by clinical disease or rise in temperature. The vector is unknown.

4. Trypanosomiosis

The disease, commonly known as 'Surra' and caused by a flagellate blood protozoan parasite, *Trypanosoma evansi*, is widely prevalent among different species of domestic livestock especially in horses, camel, cattle, buffaloes, dogs and other host species. Cattle and buffaloes are generally considered to act as reservoirs of the infection and are potential danger to other susceptible livestock (Anon, 1977). Ayyar (1920), while recording outbreak among buffaloes, observed that symptoms seen in buffaloes were similar to those of cattle and that a symptomatic carrier state might be accompanied by loss of condition and progressive anaemia. Peracute cases were characterized by high fever, signs of abdominal pain, stertorous breathing and groaning. In acute cases, there was congestion of conjunctiva, keratitis or even blindness, nasal catarrah and sometimes cerebral involvement. The disease could be mistaken for anthrax or haemorrhagic septicaemia of young and weak animals. Males were said to suffer more. According to Sheather (1923), buffaloes were only slightly susceptible and seldom showed symptoms while fatal cases of 'Surra' in buffaloes have also been reported (Bachan Singh, 1936; Sundram, 1936; Viswanathan, 1947; Singh, 1950).

Usually, infected bovine and bubaline hosts become carriers of the infection and the parasites appear in the peripheral circulation periodically and usually in scanty numbers so that they escape detection on microscopic examination of blood. Such inapparent infection, however, may flare up into clinical condition if the animals are stressed by inclement weather or also infected concomitantly with liverfluke or any other infection like rinderpest, foot and mouth, haemorrhagic septicaemia piroplasms, anaplasma, etc. Flaring up of latent 'Surra' due to immunization against rinderpest has also been reported in cattle. Slight impairment of the body defenses would result in imbalance in favour of the parasites (Gill, 1977).

Verma and Gautam (1977, 1977a, 1978, 1979), in their series of publications, studied some aspects of the pathological condition in experimental *T. evansi* infection in buffalo-calves e.g. serum protein changes, serological diagnosis, course of the disease and gross and histopathological changes. According to them, there was no change in total serum protein but there was significant increase in beta and gamma globulins with decrease in alpha globulin and albumin and because of non-specific nature of gamma globulin, the animals succumbed to the infection. They reported that 10 buffaloes, infected with a virralent strain of *T. evansi* by sub-cutaneous route, suffered either from subacute or chronic form of surra with 80 percent mortality and with a maximum period of carrier state of 190 days in a buffalo-calf. The most prominent gross pathological lesions, as reported by the two workers, were extensive gelatinisation in the pelvis of the kidneys and on the base of the heart, enlargement of superficial lymph glands, congestion and oedema of lungs, congestion of liver, pulmonary emphysema and atrophy of gluteal muscles. The blood was dark in colour and was partially coagulated. On histopathological examination, congestion, loss of normal architecture, absence of germinal follicles, thickening of trabeculae and trabecular artery, excess of plasma cells and haemosiderin deposits were obseıved in the spleen. There were degenerative changes in kidney, liver and heart. Lungs werc found congested and oedematous.

Symptoms

The acute form of 'Surra' in buffaloes may become fatal. The animal usually dies within a day or two after showing high fever and symptoms of intense nervous excitement (delirium) alternating with those of severe depression (coma). It moves aimlessly in circles and frequently fall down. Grinding of the teeth, salivatlon, groaning, stertorous breathing, abdominal pain, frequent voiding of urine and faeces followed by partial loss of senses, total loss of strength and prostration are common symptoms. Bleeding from the nose and ears may also be seen in some cases (Anon, 1977).

Trypanosoma theileri is only other trypanosome sometimes encountered in buffaloes and other hosts in India. It is much larger than *T. evansi* but, owing to its low density in the blood stream, microscopical examination often fails to detect the infection. This species is generally nonpathogenic but, under condition of stress, it may be incriminated as a cause of serious condition.

Epidemiological data on the incidence of 'Surra' in different parts of the country are lacking.

Diagnosis, in acute infection, is readily done by demonstration of organisms in freshly stained blood smears. In chronic cases, thick and thin blood smears or lymph node puncture smears should be examined more than once. Other indirect diagnostic tests include a number of chemical tests viz. mercuric chloride test, stillbamide test and others but are inconsistent (Gill, 1977). Serological test viz. complement fixation test, indirect haemagglutination test, fluorescent antibody test and others are also employed in the diagnosis of trypanosomiasis. Animal inoculation in guinea pigs and rats is more confirmatory than other methods of diagnosis.

5. Anaplasmosis

It is a specific infectious disease of ruminants characterized by variable degree of pyrexia, emaciation, anaemia and icterus. It is doubted by many workers that the causative agent, *Anaplasma* is a protozoan and evidently considered to be rickettsial in nature. Gautam and Singh (1970) reported 33.3 percent of buffaloes in Punjab and Haryana to carry the infection of *A. marginale* as intraerythrocytic rounded and/or a filamented form. Younger animals are more susceptible and severe clinical infections commonly occur in cattle. Species of ixodid ticks, tabanid flies, stable flies and mosquitoes have been demonstrated to act as vectors of this infection.

Treatment

Little information is available on the treatment of different blood protozoan diseases of buffaloes. Verma *et al* (1973) found Berenil effective in 'Surra' of cattle and buffalo, when given at the dose rate of 16 and 10 mg/kg body weight. In general, there are a number of trypanocidal drugs used in different host species viz. Antrycide, Antrypol, tartar ematic, Naganol and others (Gill, 1977). Similarly, no specific treatment for piroplasmosis (babesiosis, theileriosis) have been reported in buffaloes. A few drugs commonly employed in bovine hosts in these infections are listed below :

1. Trypan blue
2. Berenil
3. Pirevan (Babesan)
4. Diethylcarhamazine citrate
5. Nivaquine, Pamaquin, Camoquin, Resochin, etc.
6. Achromycin.
7. Roletetracycline
8. Sulphadimidine and chloroquine phosphate, etc.

Babesiosis

Earlier, trypan blue @ 100 ml of 1-2 percent solution in normal saline and acriflavine @ 20 ml of 5 percent aquous solution, given intravenously, have now been replaced largely by quinuronium sulphate and diamidines.

Pirevan (Babesan, Acapron, Piroparv, Acaprin, Piroplasmin) @ 1 ml of a 5 percent solution per 50 kg. body weight is given subcutaneously. There ma y be toxic reactions consisting of vasodilation with sweating, salivation, diarrhoea, urination and sometimes collapse and death. *Phenamidine* @ 12 mg per kg bodyweight in a 40 percent aquous solution is given subcutaneously.

Berenil (Diminazine) @ 2-3.5 mg per kg bodyweight, given by deep intramuscular infection, is used to treat all species of *Babesia.*

Diampron (Amicarbilide) @ 10 mg per kg bodyweight, given by deep intramuscular injection or subcutaneously, is quite effective against *B. Bigemina.*

Imidocarb (Carbanilide, Imidazole) is used both as a therapeutic and prophylactic drug @ 0.5-1 mg per kg bodyweight given subcutaneously. This is the only compound which eliminates *B. bovis* from carrier cases. (Soulsby, 1982).

Tropical theileriosis (*T. annulata*)

Oxytetracycline, chlortetracycline or Berenil (Diminazine) @ 10-15 mg per kg bodyweight and Steclin @ 2-4 mg per kg bodyweight, given by intramuscular injection on alternate day for a week, may prove useful (Gautam and Sharma, 1972).

Primaquine is effective against stages in red blood cells. Nivaquite @ 10-15 mg/kg bodyweight is given intramuscularly for 3-4 days in early cases under field condition.

Diethylcarbamazine, in single dose, @ 50 mg per kg bodyweight given intramuscularly, pull down parasitaemia from 15-30 percent to 0-4 percent. But for a transit anurea during initial 8-10 hr, there is no other side effect (Sharma *et al*, 1976).

Though a large number of compounds have been tried alone and in combination with claim of variable rates of effectiveness against tropical theileriosis, as yet true therapeutic properties in any of these drugs has not been found.

Immunoprophyaxis: Different procedures were experimented from time to time to immunize livestock against theileriosis. In India, Gill *et al* (1976) and Gill *et al* (1978) used infection-treatment method for immunization. They infected calves using ticks and administered chlorotetracycline orally at a dose rate of 16 mg per kg bodyweight. Eight days of medication from the beginning of the infection prevented severe reaction to the immunizing infection but allowed the development of solid resistance to homologous challenge. This suppressive activity of some tetracycline administered at the time of infection has also been studied by Jagdish *et al* (1979). According to them, rolitetracycline (Reverin, Hoechst) at a dose rate of 4 mg per kg bodyweight given intramuscularly for 4 days following infection, afforded protection against *T. annulata* infection.

In India, intensive research work is being done for developing suitable tissue culture vaccine at various centres (Sharma and Sisodia, 1975; Gill *et al*, 1976; Gill *et al*, 1976a). Immunoprophylaxis of susceptible animals by the use of irradiated organisms is also receiving much attention in the country (Singh *et al*, 1979; Samanta Ray *et al*, 1980).

Control of vector-tick is equally important in controlling theileriosis. Control of ticks by suitable acarcides should be on the front-line-defence. Side by side, efforts should also be made on biological control of the ticks. Systematic investigations on ecology and incidence of *Hyalomma* ticks, involved in the transmission of *T. annulata* in the country, should be conducted which is an essential pre-requisite for their control. Besides, control of ticks by immunization of susceptible animals against them should also be tried.

Enzootic areas of theileriosis should be defined by parasitological means and serological techniques. Role played by water buffalo in enzootics of *T. annulata* infection should be studied. Efforts on production of vaccine should be enhanced.

Trypanosomiosis

The choice of compounds commonly employed as therapeutic and prophylactic agents in trypanosomiosis in livestock, in general are : Suramin (Antrypol, Bayer, 205, Naganol etc); Pyrithidium bromide (Prothidium, RD 2801); Iso-metamidium chloride (Samorin, Trypamidium); Diminazene aceturate (Diamidine, Berenil); and Quinapyramine (Antrycide) in three forms—antrycide methylsulphate, Antrycide chloride and Antrycide prosalt (a mixture of the methylsulphate and chloride). Of these, Antrycide methylsulphate has good therapeutic value against *T. evansi* infection at a dose rate of 5 mg per kg bodyweight given as a 10 percent aquous solution subcutaneosly. It is well tolerated by bovines. Besides, a mixture of 3 parts of Antrycide methylsulphate and 2 parts of Antrycide chloride, at a dose rate of 7. 4 mg per kg bodyweight of a solution of 3.5 gm in 15 ml of water, is given subcutaneously and has good therapeutic activity with protection lasting upto 2-3 months against *T. evansi* infection.

Tartar emetic is an old drug but still used to treat surra in animals because of its cheapness. Likewise Antimosan is also used in bovine surra.

References

Anon, 1977. *Handbook of Animal Husbandry*. ICAR. New Delhi pp. 788.

Ayyar, V.V.V. 1920. *Madras Vet J.* **1** : 7.

Bachan Singh, 1936. *Indian J. Vet. Sci.* **6** : 243-250.

Bhattacharyulu, Y. ; Chaudhari, R.P. and Gill, B.S. 1975. *Parasitol.* **71** : 1-7.

Bhattacharyulu, Y., Dhar, S. and Gautam, O.P. 1972. 14th Conf. Anim. Diseases. ICAR, Mathura.

Burridge, M.J. and Odeke, G.M. 1973. *Expti. Parasit.* **34** : 257-261.

Chhabra, R.C., Baxi, K.K. and Chawla, J.S. 1972. *Indian Vet. J.* **49** : 321-325.

Dhar, S., Bhattacharyulu, Y. and Gautam, O.P. 1973. *Haryana Agric. Univ. J.* **3** : 27-30.

Dhar, S. and Gautam, O.P. 1977a, *Indian J. Anim. Sci.* **47** : 389-94.

Dhar, S. and Gautam, O.P. 1977b. *Indian. J. Anim. Sci.* **47** : 458-62.

Dhar, S. and Gautam, O.P. 1977c. *Indian J. Anim. Sci.* **47** : 566-70.

Dhar, S. and Gautam, O.P. 1977d. *Indian J. Anim. Sci.* **47**. 720-23.

Dwivedi, S.K. ; Malik, K.P. and Malhotra, M.N. 1979. *Indian Vet. J.* **56** : 333-335.

Gautam, O.P. and Sharma, R.D. 1972. All India Seminar on Blood Protozoan Diseases. Feb., 1972., Hissar.

Gautam, O.P. and Singh, B. 1970. *J. Res. HAU*, **1** : 85.

Gill, B.S. 1977. *Trypanosomes & Trypanosomiasis* of *Indian Livestock*. I.C.A.R. New Delhi. pp. 136.

Gill, B.S., Bhattacharyulu, Y. and Kaur, D 1976a, *Res. Vet. Sci.* **21** : 146-149.

Gill, B.S., Bhattacharyulu, Y. and Kaur, D. 1977. *Ann. Rech. Vet.* **8** : 285-92.

Gill, B S.; Bhattacharyulu, Y.; Kaur, D. and Singh, A. 1976b. *Nature* **264** : 355-356.

Gill, B.S.; Bhattacharyulu, Y.; Kaur, D. and Singh, A. 1978. *Inter. J. Parasit.* **8** 467-469.

Gill, B.S.; Kaur, D. and Bhattacharyulu, Y. 1974. *Bull. Int. Epizoolt.* **81** : 805-811.

Hassan, S.R. 1953. *Pakistan J. Hlth.* **3** : 59 (cited by Mohan, 1968).

Idnani, J.A. 1938. *Indian Vet. J.* **14** : 311.

Jagdish, S.; Singh, D.K.; Gautam, O.P. and Dhar, S. 1979. *Vet. Rec.* **104** : 140-142.

Mantyan, E.M. 1955. *Trudy armyan. nauch. issled. Vet. Inst.* **8** : 77-81 (abstr. V.B. **27** : 1733).

Mohan, R.N. 1968. *Diseases* and *Parasites of Buffaloes. Part III. Parasitic* and *Miscellaneous Diseases. Vet. Bull.* **38** : 735-756.

Narusu, G.V. 1940. *Indian Vet. J.* **16** : 419.

Parthasarthi, P. 1931. *Indian Vet. J.* **8** : 13-21.

Raghavachari, K. 1944. *Indian Farming.* **5** : 1-3.

Roychoudhury, G.K. and Gautam, O.P. 1979. *Trop. Anim. Hlth. & Prod.* **11** : 91-93.

Sahai, L.; Dhanda, M.R., Mohan, R.N.; Rajgopalan, V.R. and Srivastava, H.D. 1962. Research in Animal Husbandry, *Indian Cong. Agric. Res.* New Delhi.

Samanta Ray, S.N. and Bhattacharyulu, Y. and Gill, B.S. 1980. *Inter. J. Parasit.* **10** : 356:58.

Sharma, R.D. and Sisodia, R.S. 1975. Summer Institute (I.C.A.R.) Deptt. Medicine, Haryana Agricultural University, Hissar.

Sheather, A.L. 1919. *Bull. Agric. Res. Inst. Pusa. No.* **10**. *J. Comp. Path.* **32** : 223-229.

Singh, D.K.; Jagdish, S. and Gautam, O.P. 1979. *Amer. J. Vet. Res.* **40** : 767-769.

Singh, S. 1950. *Indian Vet. J.* **27** : 74-77.

Soulsby, E.J.L. 1982. *Helminth. Athropods* and *Protozoa of Domesticated Animals.* Seventh Edn. The English

Language Book Society and Bailliere Tindall. London pp. 809.
Sundram, P.S. 1936. *Indian Vet. J.* **23** . 290-299.
Verma, B.B. and Gautam, O.P. 1977. *Indian Vet. J.* **54** : 294-297.
Verma, B.B. and Gautam, O.P. 1977a. *Indian Vety J.* **54** : 809-813.
Verma, B.B. and Gautam, O.P. 1978. *Indian Vet. J.* **55** : 648-653
Verma, B.B. and Gautam, O.P. 1979. *Indian Vet. J,* **56** : 13-15.
Verma, B.B.; Gautam, O.P. and Mallick, P.D, 1973. *Vet. Rec.* **27** : 465.
Viswanathan, G.R. 1947. *Indian Vet. J.* **23** : 290-299.

4

4

ECTOPARASITES

1. Insect Pests

The common insect pests of buffaloes are mosquitoes, biting gnats, stable fly, buffalo fly, horse-fly and buffalo louse, (Chaudhuri, 1976). These pests, not only feed on their hosts, suck blood, cause irritation and annoyance, damage hides, skins, hairs and hoofs and loss of animal products, they spread certain pathogenic protozoan and microbial diseases.

Mosquitoes are small dipteran insects belonging to the family Culicidae with three genera, *Anopheles*, *Culex* and *Aedes*. They bite and suck blood and cause great annoyance which results in loss in body weight and milk production. Animals can be protected from mosquitoes by eliminating their breeding places and treating the water in nearby pools or in dairy by insecticides.

Biting gnats are minute dipteran insects belonging to the families Ceratopogonidae with the genus *Culicoides* (biting midges), Simulidae with the genus *Simulium* (black flies or buffalo gnats) and Psychodidae with the genus *Phlebotomas* (sand flies). Only females of these flies bite severely the animals and suck blood. The bites are very painful. *Culicoides* act as intermediate host of a filarial nematode, *Onchocerca gibsoni* causing worm nests or nodules subcutaneously in cattle and buffalo. *Simulium* may act as intermediate host of another species of *Onchocerca*, *O. gutturosa* causing nodulation in ligamentum nuchae of buffalo and cattle.

Stably fly and buffalo fly belong to the family Anthomyidae with the genera *Stomoxys* (stable fly) and the genus *Lyperosia* (*Siphona*) (buffalo fly). Common house fly of the genus *Musca* also belongs to this group of flies. Stable flies are common insect pest of livestock including buffaloes in India. Both sexes of the fly feed on blood by their piercing and sucking proboscis. Buffalo flies also suck considerable amount of blood of the host. A great annoyance is caused to the animal in mass attack which adversely affect the milk production. To get rid of the attack, the animals rub the affected parts of their bodies vigorously against hard objects which results in wounds. Both the flies are said to involve in the transmission of "Surra" caused by *Trypanosoma evansi*.

Horse flies belong to another dipteran family, Tabanidae. The flies are of big size and belong to three genera, *Tabanus*, *Haematopota* and *Chrysops* which are serious pests and attack farm animals including buffaloes. Only female flies suck blood and make a wound in the skin by their sharp blade-like mouth parts. The bite is very painful to the animals and the flies attack repeatedly the animals. The attacked animals become restless and move about to escape. Sometimes, animals become unmanageable and run frantically. The flies cause great disturbance to drought animals in their working and adversely affect the milk production. They also transmit trypanosomiasis (surra), anaplasmosis and authrax.

The buffalo louse, an important pest of buffaloes, is *Haematopinus tuberculatus* belonging to the order Siphunculata (sucking lice). It is quite large in size and has piercing and sucking type of mouth parts with brownish grey body. The pest is, like other lice, fully parasitic as its all developmental stages occur on the body of the animals. Their incidence is more in the northern India and occur most commonly during winter and least during summer months. They start spreading during and just after rains and reach peak during January and February. Young animals or calves suffer most. In heavy infestation, the affected animals are seen rubbing and scratching their bodies to hard objects like walls, tree stems, mangers etc. to get rid off extensive itching caused by the biting. The animals loose hairs, scaly areas develop on the body and some times bruises may be seen on the body. The affected animals loose condition, disturb feeding and milking. The coat becomes rough and shaggy. Excessive licking of the matted hair by calves may lead to formation of hair balls in the stomach. For controlling lousiness in buffaloes, use of spray or dust in 0.2% concentration of Carabaryl, an actual ingredient of Sevin (Union Carbide), has been found to be quite effective in single application (Chauhan *et al.* 1981). Other insecticides are also effective in controlling this pest.

2. Ticks

Ticks are another group of ectoparasites belonging to the class Arachnida. They are mostly ixodid or hard ticks and are usually bean or nut-shaped acarines which inbibe large quantities of blood and tissue fluids of their hosts. Besides direct annoyance to their hosts, they are responsible for the transmission of many serious/fatal microbial/protozoan diseases. The important and commonly occurring species of ticks associated with cattle and buffalo in India belong to the genera, *Hyalomma*, *Boophilus*, *Rhipicephalus* and *Haemaphysalis*. Miranpuri and Naithani (1978) have listed 27 species under various genera from Indian buffaloes. These are mostly found in the region of the udder or the scrotum and relatively soft and vascular parts of the host's body such as neck, dewlap and the eustachian. Some species prefer hairy parts of the body where these cannot be seen easily. Others prefer bare parts of the belly, flank and around anus. Unfed ticks may survive as long as more than a year. Their incidence increases with the onset of rainy season and start receding during spring.

In heavy tick infestation, there is severe irritation and the skin becomes inflamed and scaby. Secondary infection by bacteria is commonly seen. Severely infested animals, particularly calves, may develop anaemia, loss of condition, dullness and in adult animals, there is marked reduction in milk yield. This condition is known as tick worry. Also tick infested hide may lose its value. There may also be *tick paralysis* caused by the salivary secretions having a toxin introduced into the body of the animal during feeding. Besides direct annoyance, these ticks are responsible for the transmission of certain haemoprotozoan parasites causing 'Babesiosis', 'Anaplasmosis', (by the species of *Boophilus*). and 'Theileriosis' (by the species of *Hyalomma* and *Boophilus*. Species of *Rhipicephalus* and *Haemaphysalis* may also be involved in the transmission of these protozoan infections. The reported transmission of Brucellosis and Epizootic bovine abortion by ticks in bovine from abroad is yet to be investigated and confirmed in our country. Possibilities of other micro-organisms, viruses being transmitted from animals by ticks cannot be ruled out.

Control of ticks is a great problem because they use numerous mammalian and avian hosts in addition to their protected abode in rough pastures and surviving long in unfed condition. For effective iradication, several methods are suggested eg, burning and ploughing of infested pastures and grazing lands, effective drainage and treatment of infested pastures with DDT spray. For the treatment of infested animals, there are numbers of acaricidal drugs which are used as sprays dips, dust and by direct application of prepared emulsions lotions on the infested areas of the body of animals. Use of Sevin 50 WP (Union Carbide), as emulsion in linseed oil in 0.1-0.5% concentration of the actual ingredient (Carbaryl), has been found most effective in removal of ticks in few hours. Besides, spray in water in 0.2-0.5% concentration and dust in 0.5% concentration are also effective in controlling tick infestation (Bhatia *et al*) 1982).

3. Sarcoptic mange

Buffaloes usually suffer from sarcoptic mange and often become fatal in young buffaloes. The whole body may be involved and the lesions starting either at the head and neck or perineum. The hair falls out and the skin becomes folded and scaly. Later, yellowish wrinkled crusts are formed containing numerous *Sarcoptes* mites and their stages (Bachan Singh, 1937; Sen and Fletcher, 1962; Choudhuri, 1962; Srivastava and Khan, 1965). The cases are reported to have been treated successfully with organophosphorus compounds.

4. Psoroptic mange

Timoney (1924) reported it in buffaloes in India. A special form of psoroptic scabies affecting the horns of buffaloes has been observed in India (Balekundri, 1928; Sen and Fletcher, 1962). Mites attack the skin at the root of the horn and gradually work their way towards its tip burrowing through and building nests between core and the horny appendage. The latter becomes brittle and is ultimately reduced to powder. It becomes detached exposing the sensitive core. The irritation may cause the animal to strike the horn on the manger continuously (Mohan, 1968).

Alwar and Lalitha (1961) found *Dermatophagoides* mites on buffaloes.

Myiasis, particularly due to *Chrysomyia bezziana* is also not uncommon in buffaloes in India.

References

1. Alwar, V.S. and Lalitha, C.N. 1961. *Indian Vet. J.* **38** : 292-294.
2. Bachan Singh, 1937. Cited by Sen and Fletcher, 1962.
3. Balekundri, G.S. 1928. *Indian Vet. J.* **4** : 340-341.
4. Bhatia, B.B., Chauhan, P.P.S., Agarwal, R.D. and Katara, R.P. 1982. (Unpublished).
5. Chauhan, P.P.S., Bhatia B.B., Agarwal, R.D. and Katara, R.P. 1981. *Vet. Res. J.* **4** : 122-24.
6. Chaudhuri, R.P. 1962. *Proc. 49th Indian Sci. Cong. Cuttack* Pt. III (Abstr.) p. 440.
7. Chaudhuri, R.P. 1976. *Insect Pests of Farm Animals in India*, I.C.A.R., New Delhi. pp. 55.
8. Miranpuri, G.S. and Naithani, R.C. 1976. *A check list of Indian Ticks, Parasitology Division*

I.V.R.I., Izatnagar, (U.P.) p. 50.

9. Mohan, R.N. 1968. Diseases and Parasites of Buffaloes. Part III. Parasitic and Miscellaneous Diseases Vet. Bull. **38** : 735-756.

10. Sen, S.K. and Fletcher, T.B. 1962. *Veterinary Entomology and Acrology for India*. I.C.A.R., New Delhi.

11. Srivastava, S.C. and Khan, 1965. *Agric. Res. New Delhi* **5** : 288.

12. Timoney, 1924. *Bull. Agric. Res. Inst. Pusa., No.* **154** : 180-200. (Cited by Mohan, 1968).

5

MISCELLANEOUS INFECTIONS

1. Sarcocystis and Sarcocystosis

Sarcocystis species have the life-cycle proliferating in intermediate hosts leading to elongate cysts developing in muscles with gametogony and sporogony occurring in final hosts producing isosporan type of oocysts with two sporotysts, each with 4 sprozoites.

According to Babudieri (1932), Moule recorded for the first time the occurrence of *Sarcocystis* in water buffalo in 1888 without naming the species. Railliet (1897) named it *Balbiania fusiformis* which was subsequently transferred by Bernard and Bauche (1912) to the genus *Sarcocystis* created by Lankester (1882) and named as *S. fusiformis*. Till recent past, this name was used for both the forms of cattle and buffalo though Wenyon (1926) had already indicated that water buffalo was the only host for *S. fusiformis*. Levine (1977) maintained this species to its rightful host—the buffalo.

From India, *S. fusiformis* was reported by Rahimuddin (1941) in 26.7 percent in 600 buffaloes in South India. Ramanujachari and Alwar (1951) found it in 10 percent of buffaloes. Chauhan *et al* (1978) observed an overall incidence of *S. fusiformis* in 9.37 percent buffaloe in the northern part of the country and recorded its occurrence in ocular musculature in addition to its usual sites in the oesophagus and tongue.

Chauhan *et al* (1977) were the first to recover sporulated sporocyst-like cysts in the faeces of dogs after feeding *S. fusiformis* (?) cysts from buffaloes. Simultaneously, Dissanaike *et al* (1977) incriminated the cat as the definitive host of *S. fusiformis*. Dissanaike and Kan (1978) described another species of *Sarcocystis* from the buffalo and named it *S. levinei*. Later, Ghosal and Shah (1979) proved itsv alidity and reported 80 per cent of buffaloes carrying both the species viz. *S. fusiformis* and *S. levinei*—the former having cat and the latter having dog as their final hosts. Jain (1982) reported gamestogonous and sporogonous stages of *S. levinei* studied in 13 dogs and of *S. fusiformis* in 10 cats and described macro- and microgamonts and oocysts in different stages of sporogony in the intestinal lining.

Railliet (1891) first described sacrocysts in muscles as white, fusiformis bodies with one end more pointed than the other measuring 5.0 to 15.0 mm by 2.0 to 4.0 mm as macroscopic. Wettimuny and Abeysene (1966) stated that sarcocysts in cattle and buffalo could easily be seen with naked eye as thin, white streaks running parallel to the longitudinal axis of the host muscles. These, in the buffalo, measured 0.5 cm to 3.0 cm. Retzlaff and Weise (1969) first pointed out two kinds of sarcocysts in muscles of naturally infected buffaloes in Turkey and called them 'macrocysts' and microcysts. The macrocysts were white spindle-shaped with one end slightly rounded and the other pointed and 7.0 mm to 25.0 mm long. The sarcocyst cavity was divided into numerous compartments by septa. Dissanaike *et al* (1977) also found two types of sarcocysts in the oesophageal muscles of buffaloes in Malaysia—the large ones, visible by naked eye, assigned to *S. fusiformis* utilizing the cat as its definitive host and the other microscopic one, later named as *S. levinei* (Dissanaike and Kan, 1978), utilising dog as its definitive host. Ghosal (1978) described *S. fusiformis* as white spindle shaped bodies in muscles of buffalo. Ghaffar *et al* (1978), from Egypt, also described two types of sarcocysts as large and small from the oesophageal muscles of naturally infected buffaloes. The large ones measured 7.0 to 30 mm by 0.7 to 2.0 mm in size while small ones measured 1.3 to 5.0 mm by 0.7 to 2.0 mm in size.

Based on the information available so far, it is evident that buffalo acts as intermediate for two distinct species of *Sarcorocystis*, *S. fusiformis* (large size) and *S. levinei* (small size) with cat and dog as their final hosts respectively. Both the species are widely prevalent in buffaloes in India but there is no report either on the clinical signs or the gross lesions due to sarcocystosis in this host. On the other hand, *S. cruzi* is proved to be highly pathogenic to cattle with a variety of symptoms including abortions (Dubey, 1976; Dubey and Bergeron, 1982). It is likely that sarcocysts in buffaloes too may be responsible for certain losses including problem of calf mortality attributed to unknown causes. Pathogenic aspects, therefore, need detailed investigation.

2. Infection with bladder worms (cysticerci and hydatid cysts) are commonly seen in buffaloes. There are a number of published and unpublished reports on the occurrence of *Cysticercus tenuicollis* and *C. bovis* in buffaloes in northern India (Thapar, 1956). Gill and Rao (1967) reported hydatid infection in 24 percent of buffaloes in Bareilly. Pathogenicity varies with the

intensity of infection.

3. Among other infections, Dwivedi and Singh (1966) saw amoeba-like organisms, thought to be *Acanthoamoeba* in necrotic foci in lungs of a buffalo affected with nodules. Shalash (1961) alone reported trichomoniasis in buffalo.

References

1. Babudieri, B. 1932. *Arch. Protistenk.* **76** : 421-580 (Cited by Levine, 1977).
2. Bernard, P.N. and Bauche, J. 1912. *Bull. Soc. Path. Exot.* **5** : 109-114. (Cited by Levine, 1977).
3. Chauhan, P.P.S., Agarwal, R.D. and Arora, G.S. 1978. *Indian J. Parasit.* **2** : 123-124.
4. Chauhan, P.P.S., Bhatia, B.B., Agrawal, R.D., Katara, R.P. and Ahluwalia, S.S. 1977. *Indian J. Exptl. Biol.* **15** : 492-494.
5. Dissannaike, A.S. and Kan, S.P. 1978. *Z. Parasitankde.* **55** : 127-138.
6. Dissannaike, Kan, S.P., Retnasabapathy, A. and Baskarana, G. 1977. *Trans. R. Soc. Trop. Med. Hyg.* **71** : 271.
7. Dubey, J.P. 1976. *Amer. J. Vet. Med. Assoc.* **169** : 1061-1078.
8. Dubey, J.P. and Bergeron, J.A. 1982. *Vet. Pathol.* **19** : 315-318,
9. Dwivedi, J.N. and Singh, C.M. 1966. *Indian J. Microbiol.* **5** : 31-34.
10. Ghaffar, F.A., Hilalim, M. Scholtyseck, E. 1978. *Tropen. Med. Parasit.* **29** : 289-294.
11. Ghosal, S.B. 1978. Ph. D. Thesis, J.N. KVV, Jabalpur.
12. Ghosal, S.B. and Shah, H.L. 1979. Proceeding 2nd Congress of Parasitol. Sec. II, 3-4.
13. Gill, H.S. and Rao, B.V. 1967. *J. Helm.* **41** : 111-114.

14. Jain, P.C. 1982. Ph. D. Thesis, Deptt. Parasitology, Coll. Vet. Sci. and A.H., Jabalpur (Unpublished)
15. Lankester, E.R. 1882. *J. Microsc. Sci.* **22** : 53-65 (Cited by Levine, 1977).
16. Levine, N.D. 1977. *J. Parasit.* **63** : 36-51.
17. Rahimuddin, M. 1941. *Indian Vet. J.* **18** : 108-109.
18. Railliet, A. 1897. *Bull. Mem. Soc. Centr. Med. Vet.* **51** : 371-377 (Cited by Levine, 1977).
19. Ramanujachari, G. and Alwar, V.S. 1951. *Indian Vet. J.* **27** : 264-266.
20. Retzlaff, V. N. and Weise, E. 1969. *Berl. Munch. Tierarztl. Wschr.* **15** : 283-286.
21. Shalash, M.R. 1961. Proc. 3rd. Wld. Congr. Fertil. Amsterdam. (Cited by Mohan, R.N. 1968. *Vety. Bull.* **38** : 735-756).
22. Thapar, G.S. 1956. *Indian J. Vet. Sci.* **26** : 211-271.
23. Wenyon, C.M. 1926. *Protozoalogy* Vol. I. Bailliere Tindal and Cox. London, pp. 778.
24. Wettimuny, S.G. and Abeysene, F.A. 1966. *Ceylon Vet. J.* **14** : 2-6.

14. [illegible] K. C. 1962, Ph. D. Thesis, Delhi [illegible]
15. [illegible] G. N., [illegible] S. and A. H. [illegible] (to be published)
16. [illegible] 2 [illegible] 22 [illegible]
17. [illegible] 1957 J. [illegible] 63 [illegible]
18. [illegible] M. 1944 [illegible] 18 [illegible] 108 [illegible]
19. [illegible] [illegible] London [illegible]
20. Ramachandran G. N. and Weiss [illegible]
21. Sharma M. R. [illegible] Ph. D. [illegible]
22. [illegible] 1956 India [illegible]
23. [illegible] [illegible]
24. [illegible] [illegible] 14 [illegible] 246

6

GENERAL CONTROL MEASURES

Control and prophylaxis of different types of parasitic infections generally becomes difficult mainly due to the following reasons : absence of characteristic clinical manifestations and/or their overlapping due to concurrent infections; absence of stages like eggs, larvae, microfiliariae or cysts being passed or shed by the host; the acute phase of the disease; lack of information on the incidence and extent of prevalence of commonly occurring parasites in the operational area and other epidemiological aspects; and proper diagnosis in time. The general principles of control of parasitic diseases are :

Protection of young or susceptible animals

The first line of defence generally considered in parasitic diseases is prevention which would include quarantine, environmental hygiene, education and protection from other infectious diseases by mass immunization which make the animals more to suffer from parasitic infections. Protection from infections may be adopted by (1) keeping the young and newly born animals away from the older stock for carrier or infected cases both in the shed and on grazing land; avoiding the water collections, ponds, ditches, etc harbouring snails and infected pasture areas for taking the animals for grazing and watering; adopting hygienic measures in the shed; providing better nutrients and clean water; and by regular prophylatic dosing with anthelmintics and other antiparasitic drugs after frequent check-up.

To prevent the animals from acquiring fluke infection like liver-fluke, amphistomes and schistosomes, snails, acting as intermediate hosts, should be eradicated. Eradication of snail population in infected areas is advocated usually by three methods—biological, mechanical and chemical. The choice depends upon local conditions and resources available. Biological control of snails is done by propagating the oligochaetes, some aquatic plants, fungi, algae, giant water bugs and by duck rearing. Mechanical removal of snails from water collections is rather difficult. In this, extermination of aquatic snails, removal of vegetations and drainage of water may be done. Chemically, snails are killed by certain chemicals eg. copper pentach lorophenate, sodium pentaclorophenate, Frescon, Bayluscide, SRC 4402 etc. The best agent, at present, for destroying snails is N-tritylmorpholine (Montague, 1967) and is sold under the trade name of Frescon which has the advantage of not being poisonous to stock over the old agent, copper sulphate. It should be applied on the farms only in the wet areas where snails thrive and in a period of wet wheather at a time when snails are multiplying. It is best applied when there is slight rain which may help to hold it in the soil without washing it away. The application of N-tritylmorpholine at 0.45 kg per hectare results in a 99 percent mortality of *Lymnaea truncatula*, a mudsnail, at high elevations acting as intermediate host for *Fasciola hepatica* (Othman and Baker, 1981). It can be tried in India with success. These chemicals should be used judiciously to avoid cases of poisoning among animals and human population visiting areas of treatment. It is better to avoid such snail-infested ponds, ditches and water collections for taking the animals for watering or wallowing.

There are number of blood protozoan infections as also filarial nematodes which are transmitted by a number of insects and archnids. To protect the animals from these infections, breeding of mosquitoes and flies can be avoided by keeping the surroundings clean, disinfected, dry and by frequent removal of dung etc. Mosquitoes and flies, if present, may be removed by fumigation, smoke and by use of repellents on the body of the animals. Further, ticks, acting as vector for priplasms, should not be allowed to stick on the bodies of animals and these should be removed and destroyed mechanically and by frequent use of effective acaricides.

Treatment of all clinical, subclinical and carrier cases of parasitosis and parasitiasis should be done in time after proper diagnosis by clinical symptoms, faecal examination, examination of nasal and eye secretions, larval culture, recovery and identifications of larvae by Bayerman's technique in living condition of the animals. In case some animals have died, immediate postmortem examination may be done by gross-examination and also histopathological examination. Immuno-diagnosis is also heading way in parasitic infections in buffaloes.

Immuno-prophylaxis in control of parasitic infections in buffaloes is yet to develop. Research on the possibility of immune prophylaxis particularly against fasciolosis and theileriosis is under active investigation at several laboratories in the world including some in India. Dargie *et al.* (1974) found that fluke burdens of calves previously immunized with metacercariale attenuated by X-rays were significantly reduced (71 percent) as compared with those of non-immunized control.

References

1. Dargie, J.D., Armour, J., Rushton, B. and Murray. 1974. *Proc. 6th Congr. Wld. Assoc. Adv Vet. Parasit.* Vienna. 1973.
2. Montague, J. Th. W. 1967. Proc. 2nd. Inter. Liver Fluke Collog. Wageningen, Netherland.
3. Othman, A.M. and Baker, S.A. 1981. "Prevention and control of common buffalo diseases in *Recent Advances in Buffalo Research and Development*. FFTC BOOK series No. 22, Taiwan, 1981, pp. 205.

7

LIST OF PARASITES OF BUFFALOES

TREMATODES

Liver

1. *Fasciola gigantica*
2. *Fasciola hepatica*
3. *Dicrocoelium dendriticum*
4. *Eurytrema pancreaticum*
5. *Gigantocotyle explanatum*

Digestive tract

1. *Paramphistomum cervi*
2. *Paramphistomum gotoi*
3. *Paramphistomum microbothrium*
4. *Cotylophoron cotylophorum*
5. *Calicophoron calicophorum*
6. *Carmyerius gregarius*

7. *Fischoederius elongatus*
8. *Fischoederius cobboldi*
9. *Gigantocotyle explanatum*
10. *Gastrothytax crumenifer*
11. *Ceylonocotyle streptocoelium*
12. *Schistosoma spindale*
13. *Schistosoma indicum*

Circulatory system

1. *Schistosoma spindale*
2. *Schistosoma indicum*
3. *Schistosoma nasalis*
4. *Ornithobilharzia turkestanicum*

NEMATODES

Digestive tract

1. *Neoascaris vitulorum* (*Toxocara vitulorum*)
2. *Haemonchus contortus*
3. *Haemonchus bubalis*
4. *Haemonchus similis*
5. *Haemonchus bedfordi*
6. *Mecistocirrus digitatus*
7. *Trichostrongylus exei*
8. *Strongyloides papillosus*
9. *Trichostrorgylus colubriformis*
10. *Paracooperia nodulosa*
11. *Srkjbinagia boevi*
12. *Ostertagia ostertagi*
13. *Cooperia onchophora*
14. *Cooperia punctata*
15. *Cooperia pectinata*
16. *Bunostomum phlebotomum*
17. *Gaigeria pachyscelis*

18. *Oesophagostomum radiatum*
19. *Chabertia ovina*
20. *Trichuris ovis*
21. *Trichuris discolor*
22. *Gongylonema pulchrum*

Circulatory system

1. *Onchocerca armillata*
2. *Elacophosa poeli*

Eyes

1. *The'azia rhodesii*
2. *Thelazia skrjabini*
3. *Thelazia gulosa*
4. *Parafilaria bovicola* (immature stages)

Peritoneal cavity

1. *Seteria labiato-papillosa*
2. *Seteria digitata*

Muscle, tendons, ligaments, connective tissues and skin

1. *Onchocerca gutturosa*
2. *Onchocerca gibsoni*
3. *Stephanofilaria assamensis*
4. *Parafilaria booicola*
5. *Onchocerca armillata* (microfilariae)

Ear

1. *Stephanofilaria zaheeri*

ratory system

1. *Dicytocaulus viviparous*
2. *Syngamus laryngeus*

CESTODES

Digestive tract

1. *Moniezia expansa*
2. *Moniezia benedeni*
3. *Avitellina centripunctata*
4. *Thysaniezia giardi*

Liver, muscles and Peritoneal cavity

1. *Cysticercus tenuicollis*
2. *Cysticercus bovis*
3. *Hydatid cysts*

Respiratory system

1. *Hydatid cysts*

PROTOZOA

Digestive tract

1. *Eimeria auburnensis*
2. *Eimeria bareillyi*
3. *Eimieria bovis*
4. *Eimeria bukidnonensis*
5. *Eimeria braziliensis*
6. *Eimeria canadensis*
7. *Eimeria cylindrica*
8. *Eimeria ellipsoidalis*

9. *Eimeria gokaki*
10. *Eimeria subspeherica*
11. *Eimeria alabamensis*
12. *Eimeria wyomingensis*
13. *Eimeria zuernii*

Circulatoiy system

1. *Babesia bigemina*
2. *Plasmodium bubalis*
3. *Theileria annulata*
4. *Theileria mutans*
5. *Trypanosoma evansi*
6. *Anaplasma marginale*

Muscles

1. *Sarcocystis fusiformis*
2. *Sarcocystis levenei*

ARTHROPODS

1. *Haematopinus tuberculatus*
2. *Haematopinus eurysternus*
3. *Simulium percuarum*
4. *Haematobia exigua* (*Lyperosia exigua*)
5. *Chrysomyia bezziana*
6. *Boophilus decoloratus*
7. *Boophulus annulatus*
8. *Boophulus microplus*
9. *Hyalomma anatolicum anatilicum* (*H. excavatum*)
10. *Hyalomma detritum*
11. *Hyalomma dromedarii*
12. *Haemophysalis bispinosa*
13. *Sarcoptes scabiei*
14. *Psoroptes natalensis*

Index

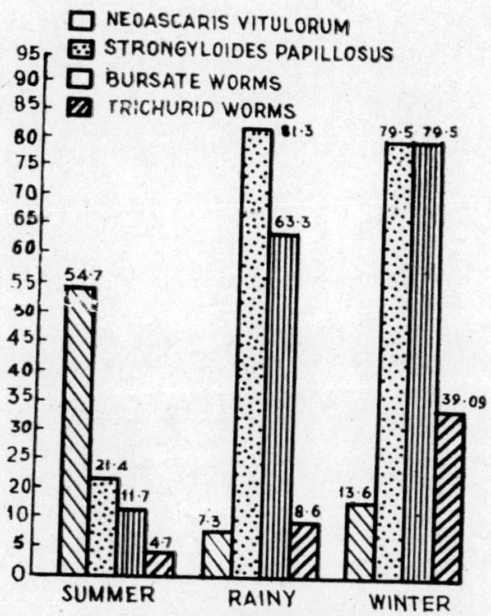

Fig. 1

Seasonal incidence of 4 types of G.I. nematodes in Buff. Calves.

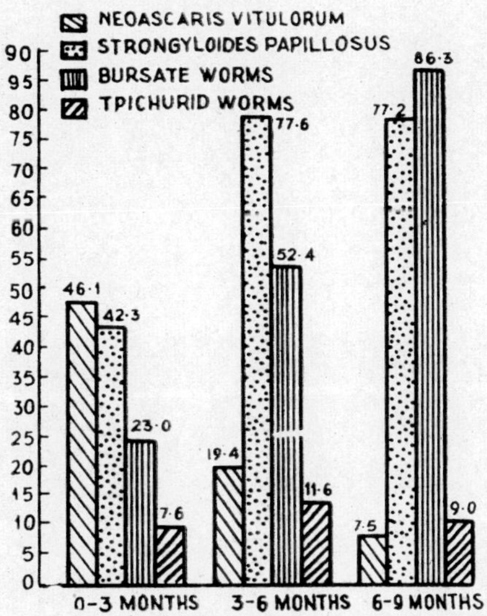

Fig. 2

Age-wise incidence of 4 types of G.I. Nematodes in Buffalo Calves

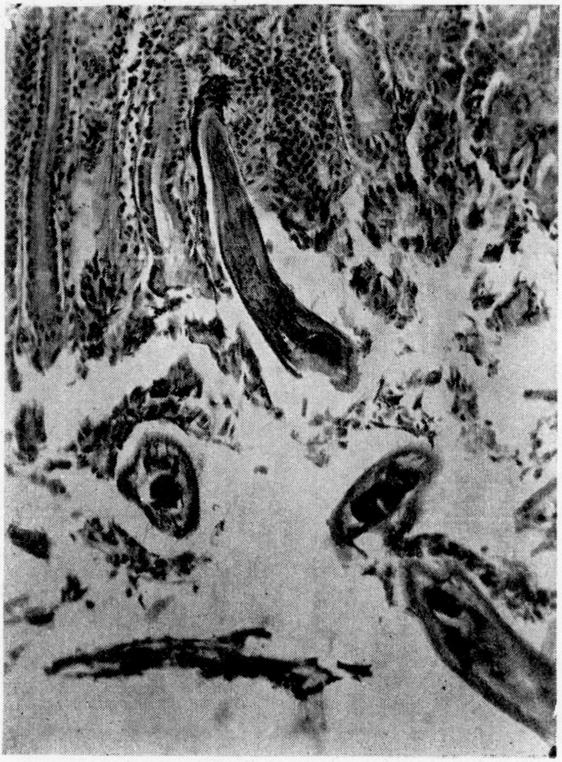

Fig. 3

Skrjabinagia boevi *in abomasal mucosa.*

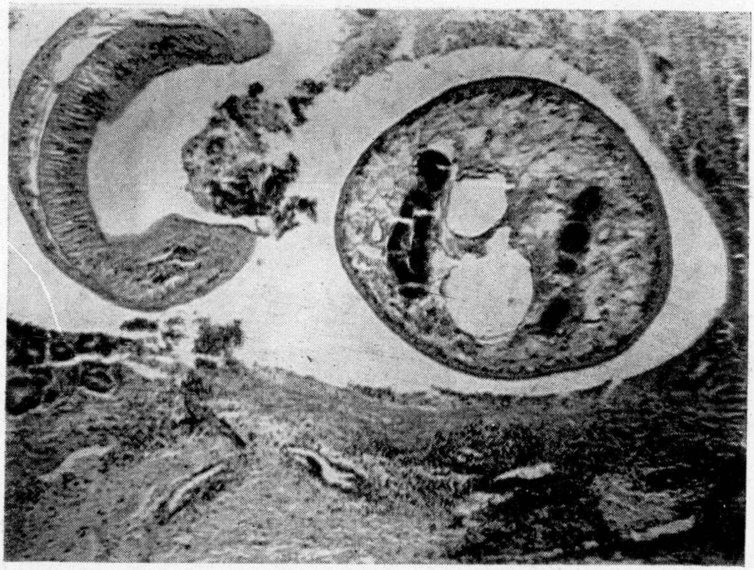

Fig. 4

Immature **Paramphistomum (Gigantocotyle) explanatum** *flukes attached to the duodenal wall,*

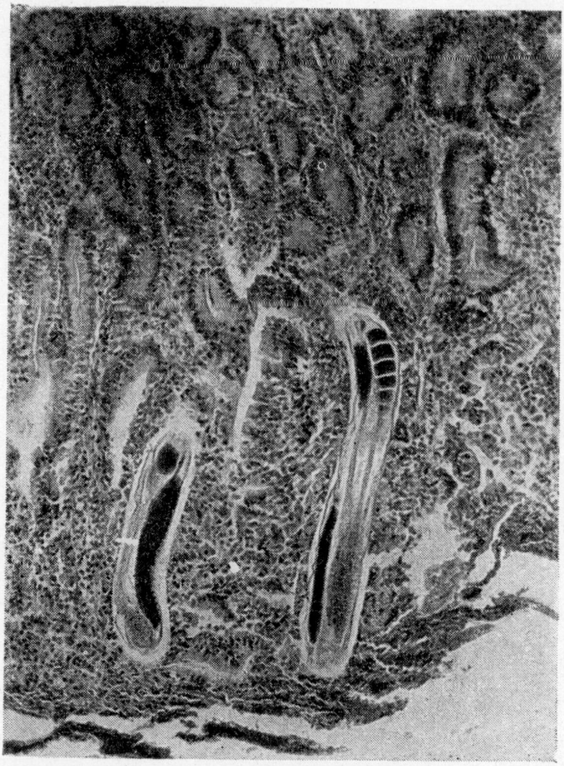

Fig. 5

Strongyloides papillosus *female worms in the small intestine.*

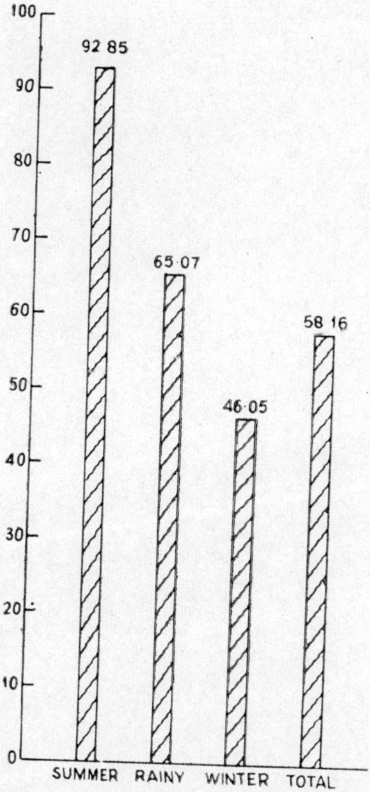

Fig. 6

Seasonal incidence of **Paracooperia Nodulosa**
in Buffalo-calves in three different seasons.
(Positivity by presence of both Nodules and free worms)

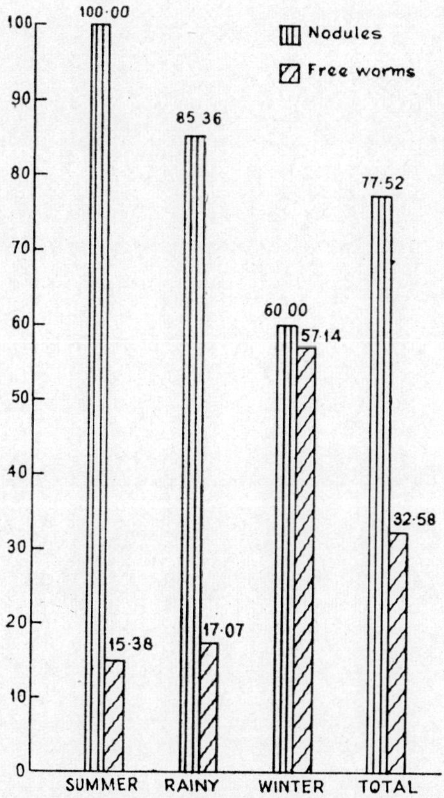

Fig. 7

Seasonal distribution of nodules and intensity of free worms in intestine of buffalo-calves, found positive for **Paracooperia Nodulosa** *infection, in three different seasons.*

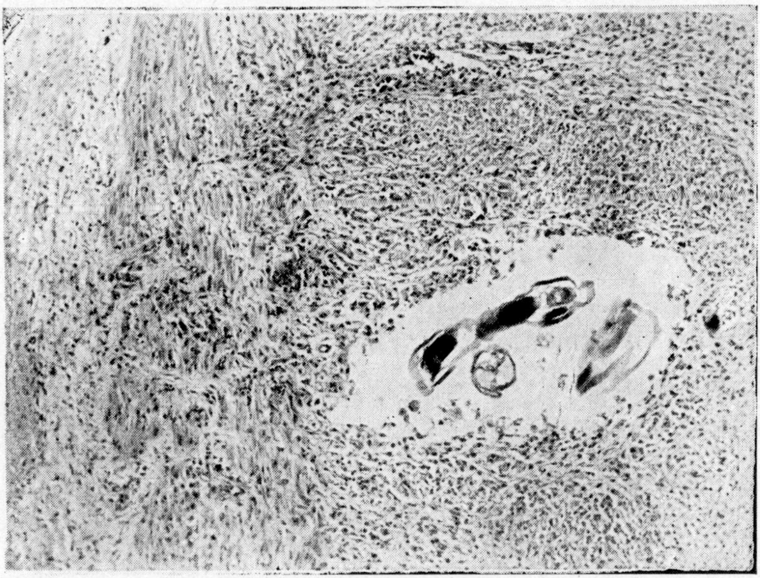

Fig. 8-1

Juveniles of **Paracooperia nodulosa** *cut in an early nodule in small intestine.*

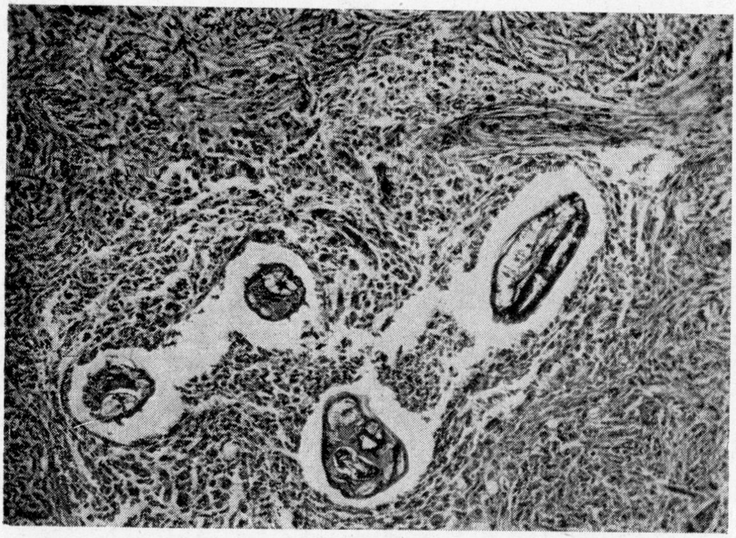

Fig. 8-2

Juveniles of **Paracooperia nodulosa** *cut inside an older nodule with fibrosis.*

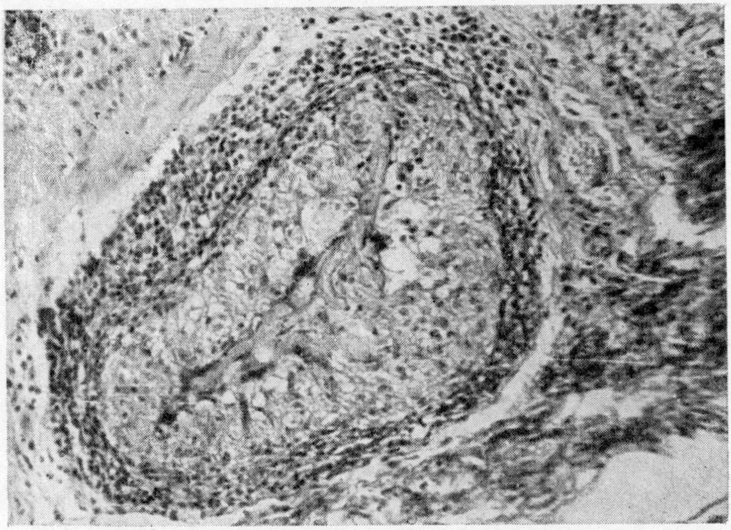

Fig. 9

Phagocytised egg of **Schistosoma spindale** *in a pseudo-tubercle in the intestinal wall.*

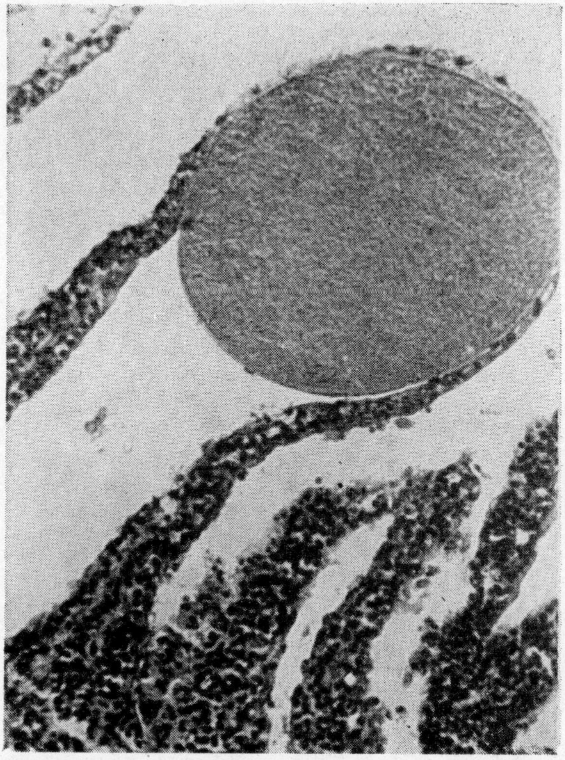

Fig. 10

A giant eimerian schizont cut in the mucosa of small intestine.

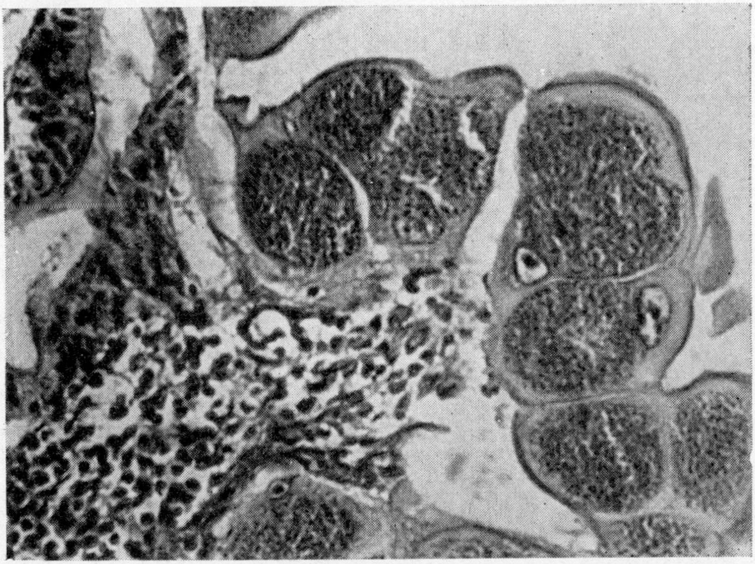

Fig. 11

Large epithelial eimerian schizonts cut in the mucosa of small intestine.

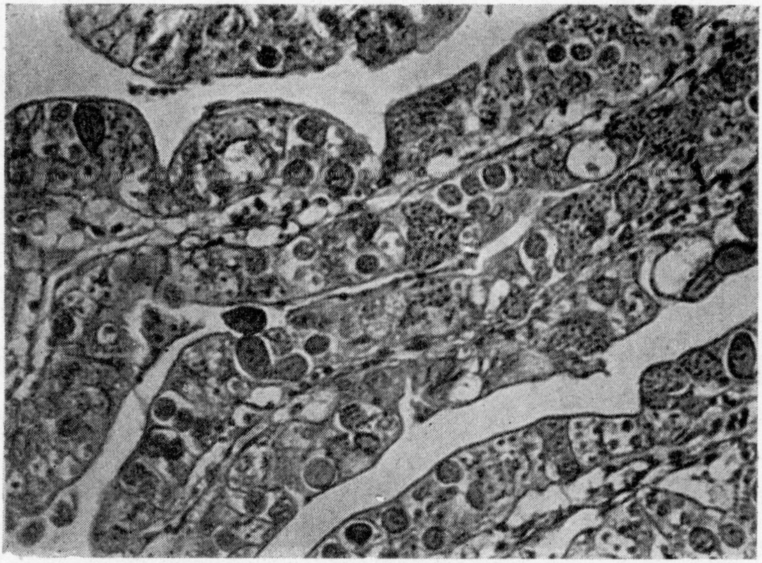

Fig. 12

Gametocytes and oocysts of **Eimeria bareillyi** *in the mucosa of small intestine.*

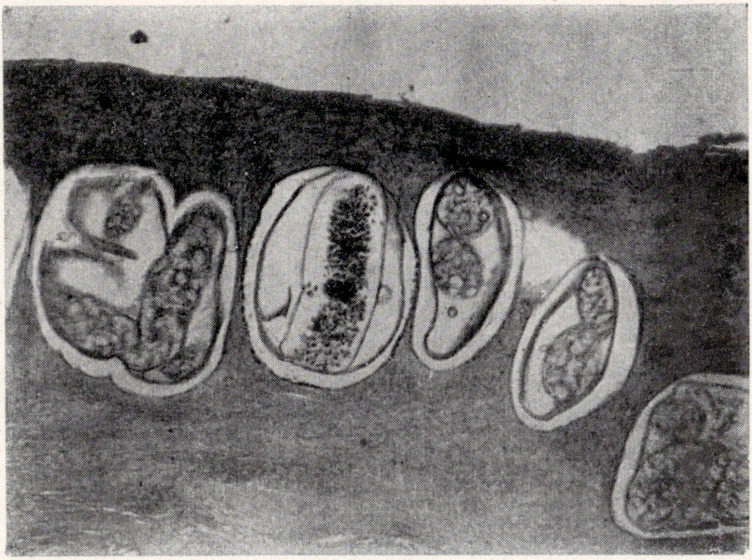

Fig. 13

Onchocerca armillata *female worms cut inside tunnels in aortic wall.*

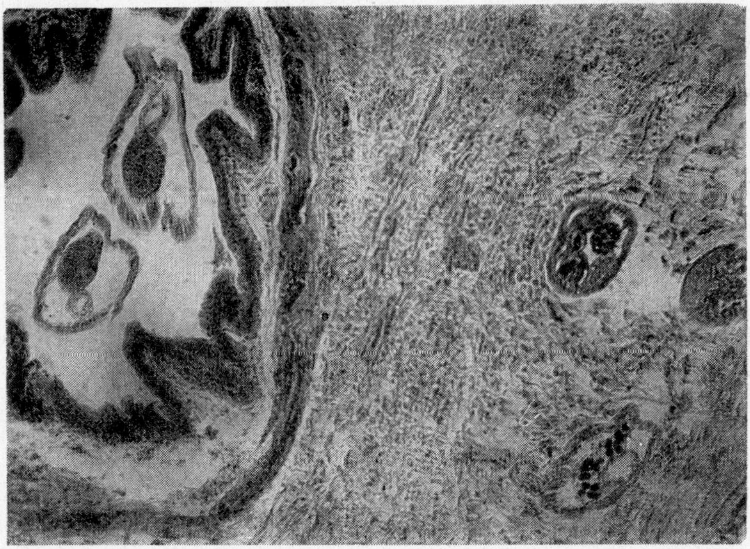

Fig. 14

Thelazia skrjabini *eyeworms cut inside lacrimal duct and connective tissues of eye ball.*